T0234729

SpringerBriefs in Research and Innovation Governance

Editors-in-Chief

Doris Schroeder, Centre for Professional Ethics, University of Central Lancashire, Preston, Lancashire, UK

Konstantinos Iatridis, School of Management, University of Bath, Bath, UK

SpringerBriefs in Research and Innovation Governance present concise summaries of cutting-edge research and practical applications across a wide spectrum of governance activities that are shaped and informed by, and in turn impact research and innovation, with fast turnaround time to publication. Featuring compact volumes of 50 to 125 pages, the series covers a range of content from professional to academic. Monographs of new material are considered for the SpringerBriefs in Research and Innovation Governance series. Typical topics might include: a timely report of state-of-the-art analytical techniques, a bridge between new research results, as published in journal articles and a contextual literature review, a snapshot of a hot or emerging topic, an in-depth case study or technical example, a presentation of core concepts that students and practitioners must understand in order to make independent contributions, best practices or protocols to be followed, a series of short case studies/debates highlighting a specific angle. SpringerBriefs in Research and Innovation Governance allow authors to present their ideas and readers to absorb them with minimal time investment. Both solicited and unsolicited manuscripts are considered for publication.

More information about this series at http://www.springer.com/series/13811

Bernd Carsten Stahl

Artificial Intelligence for a Better Future

An Ecosystem Perspective on the Ethics of AI and Emerging Digital Technologies

Foreword by Julian Kinderlerer

 Springer

Bernd Carsten Stahl
Centre for Computing and Social Responsibility
De Montfort University
Leicester, UK

Horizon 2020 Framework Programme The book will be primarily based on work that was and is undertaken in the SHERPA project (www.project-sherpa.eu). It can, however, draw on significant amounts of work undertaken in other projects, notably Responsible-Industry (www.responsible-industry.eu), ORBIT (www.orbit-rri.org) and the Human Brain Project (www.humanbrainproject.eu). The SHERPA consortium has 11 partners from six European countries (representing academia, industry, civil society, standards bodies, ethics committees, art).

ISSN 2452-0519 ISSN 2452-0527 (electronic)
SpringerBriefs in Research and Innovation Governance
ISBN 978-3-030-69977-2 ISBN 978-3-030-69978-9 (eBook)
https://doi.org/10.1007/978-3-030-69978-9

This Springer imprint is published by the registered company Springer Nature Switzerland AG
The registered company address is: Gewerbestrasse 11, 6330 Cham, Switzerland

Foreword

The volume of data collected daily about each of us, as we use the internet and social media, is immense. That data can be used in all sorts of ways, from tracking our behaviour to ensuring that advertising and information are tailored for specific individuals. Collated data may also be used to provide the raw material for Artificial Intelligence (AI) systems. Computers have become ubiquitous and are used to control or operate all manner of everyday items in ways that were unimaginable only a few years ago. Smart phones, able to track where we are and who we meet, are commonplace. Autonomous weapons capable of deciding independently what to attack and when are already available to governments—and, by extension, to terrorists. Digital trading systems are being used to rapidly influence financial markets, with just 10% of trading volume now coming from human discretionary investors (Kolakowski 2019). AI systems can (and are) being used to redefine work, replacing humans "with smart technology in difficult, dirty, dull or dangerous work" (EGE 2018: 8). The loss of jobs is likely to become a major factor in what is now termed the "post-industrial society". New jobs and new opportunities for humans need to be created. In medicine, AI is assisting in the diagnosis of illness and disease, in the design of new drugs and in providing support and care to those suffering ill health.

In many instances, AI remains under the control of users and designers, but in increasing numbers of applications, the behaviour of a system cannot be predicted by those involved in its design and application. Information is fed into a "black box" whose output may affect many people going about their daily lives:

> Without direct human intervention and control from outside, smart systems today conduct dialogues with customers in online call-centres, steer robot hands to pick and manipulate objects accurately and incessantly, buy and sell stock at large quantities in milliseconds, direct cars to swerve or brake and prevent a collision, classify persons and their behaviour, or impose fines. (EGE 2018: 6)

Newly developed machines are able to teach themselves and even to collect data. Facial recognition systems scan crowds as they make their way through the streets to detect presumed troublemakers or miscreants.

We need to ensure that the values we hold as a society are built into the systems we take into use, systems which will inevitably change our lives and those of our children. The Charter of Fundamental Rights of the European Union delineates the values that society wishes to see implemented. Those designing these systems and drafting the algorithms that drive them need to be aware of the ethical principles that underlie society. Margaret Thatcher once said that "there's no such thing as society" (Thatcher 2013), but rather there were individuals. The rise of AI is changing that, as we become identifiable ciphers within the big data used for AI. AI systems have to ensure the safety and security of citizens, and provide the safeguards enshrined in the Charter.

Control of big data, and of the AI revolution, is in the hands of a small group of super-national (or multinational) companies that may or may not respect the rights of people as they use our information for commercial or political purposes.

The advent of AI has given much to society, and ought to be a force for good. This book therefore comes at an important juncture in its development. Bernd Stahl leads a major project, SHERPA (Shaping the Ethical Dimensions of Smart Information Systems), which has analysed how AI and big data analytics impact ethics and human rights. The recommendations and ideas Bernd puts forward in this book are thought-provoking – and it is crucial that we all think about the issues raised by the impact of AI on our society.

These are exciting times!

Sheffield, England   Julian Kinderlerer

References

EGE (2018) European Group on Ethics in Science and New Technologies: Statement on artificial intelligence, robotics and "autonomous" systems. European Commission, Brussels. https://doi.org/10.2777/531856

Kolakowski M (2019) How robots rule the stock market (SPX, DJIA). Investopedia, 25 June. https://www.investopedia.com/news/how-robots-rule-stock-market-spx-djia/. Accessed 10 Nov 2020

Thatcher M (2013) Margaret Thatcher: a life in quotes. The Guardian, 8 April. https://www.theguardian.com/politics/2013/apr/08/margaret-thatcher-quotes. Accessed 10 Nov 2020

Emeritus Prof. Julian Kinderlerer   is a visiting Professor in the School of Law at the University of KwaZulu-Natal, Emeritus Professor of Intellectual Property Law at the University of Cape Town and former Professor of Biotechnology and Society at Delft University of Technology. He is the immediate past president of the European Group on Ethics in Science and New Technologies (EGE), which advises the European Commission, Council and Parliament on ethical issues. He has acted as a Director at the United Nations Environment Programme providing guidance to countries on legislation and regulation for the use of living modified organisms, and was a member of the advisory board for the SHERPA project.

Acknowledgements

This book could not have come into existence without the contributions and support of many groups and individuals in a range of projects. I would like to thank everybody who has contributed to these projects and with whom I have collaborated over the years.

Particular thanks are due to the members of the SHERPA consortium for doing much of the work that informs this book. I furthermore owe thanks to contributors and collaborators from other projects, notably the Human Brain Project, the Responsible-Industry project, the CONSIDER project and the ETICA project. Special thanks are due to Doris Schroeder, who supported the development of this book, not only as a project colleague but also as the series editor, and facilitated its publication by Springer.

The book furthermore owes much to input from colleagues at De Montfort University, in particular the members of the Centre for Computing and Social Responsibility. I hope this volume contributes to the rich research record that the Centre has established since 1995.

The book has been expertly and speedily copy-edited by Paul Wise, a brilliant editor in South Africa. I also want to thank Juliana Pitanguy at Springer for overseeing the publishing process.

My final thanks go to my family, who allowed me to lock myself away even more than legally required during the pandemic-induced time of social isolation in which the book was written.

This research received funding from the European Union's Horizon 2020 Framework Programme for Research and Innovation under Grant Agreements No. 786641 (SHERPA), No. 785907 (Human Brain Project SGA2) and No. 945539 (Human Brain Project SGA3), and the Framework Partnership Agreement No. 650003 (Human Brain Project FPA).

Contents

Chapter 1
Introduction

Abstract The introductory chapter describes the motivation behind this book and provides a brief outline of the main argument. The book offers a novel categorisation of artificial intelligence that lends itself to a classification of ethical and human rights issues raised by AI technologies. It offers an ethical approach based on the concept of human flourishing. Following a review of currently discussed ways of addressing and mitigating ethical issues, the book analyses the metaphor of AI ecosystems. Taking the ecosystems metaphor seriously allows the identification of requirements that mitigation measures need to fulfil. On the basis of these requirements the book offers a set of recommendations that allow AI ecosystems to be shaped in ways that promote human flourishing.

Keywords Artificial intelligence · Ethics · Human flourishing · Mitigation strategies · Innovation ecosystem

Artificial intelligence (AI) raises ethical concerns. Such concerns need to be addressed. These two statements are not too contentious. What is less clear is what exactly constitutes the ethical concerns, why they are of an ethical nature, who should address them and how they are to be dealt with.

AI is increasingly ubiquitous and therefore the consequences of its use can be observed in many different aspects of life. AI has many positive effects and produces social benefits. Applications of AI can improve living conditions and health, facilitate justice, create wealth, bolster public safety and mitigate the impact of human activities on the environment and the climate (Montreal Declaration 2018). AI is a tool that can help people do their jobs faster and better, thereby creating many benefits. But, beyond this, AI can also facilitate new tasks, for example by analysing research data at an unprecedented scale, thereby creating the expectation of new scientific insights which can lead to benefits in all aspects of life.

These benefits need to be balanced against possible downsides and ethical concerns. There are many prominent examples. Algorithmic biases and the resulting discrimination raise concerns that people are disadvantaged for reasons they should not be, for instance by giving higher credit limits to men than to women (Condliffe

B. C. Stahl, *Artificial Intelligence for a Better Future*,
SpringerBriefs in Research and Innovation Governance,
https://doi.org/10.1007/978-3-030-69978-9_1

2019), by referring white people more often than Black people to improved care schemes in hospitals (Ledford 2019) or by advertising high-income jobs more often to men than to women (Cossins 2018). AI can be used to predict sexual preferences with a high degree of certainty based on facial recognition (The Economist 2017), thereby enabling serious privacy breaches.

The range of concerns goes beyond the immediate effects of AI on individuals. AI can influence processes and structures that society relies upon. For example, there is evidence to suggest that AI can be used to exert political influence and skew elections by targeting susceptible audiences with misleading messages (Isaak and Hanna 2018). People are worried about losing their livelihoods because their jobs could be automated. Big multinational companies use AI to assemble incredible wealth and market power which can then be translated into unchecked political influence (Zuboff 2019).

A further set of concerns goes beyond social impact and refers to the question of what AI could do to humans in general. There are fears of AI becoming conscious and more intelligent than humans, and even jeopardising humanity as a species. These are just some of the prominent issues that are hotly debated and that we will return to in the course of this book.

In addition to the many concerns about AI there are numerous ways of addressing these issues which require attention and input from many stakeholders. These range from international bodies such as the United Nations (UN) and the Organization for Economic Cooperation and Development (OECD) to national parliaments and governments, as well as industry groups, individual companies, professional bodies and individuals in their roles as technical specialists, technology users or citizens. As a consequence, discussion of the ethics of AI is highly complex and convoluted. It is difficult to see how priorities can be set and mitigation strategies put in place to ensure that the most significant ethical issues are addressed. The current state of the AI ethics debate can be described as a cacophony of voices where those who shout loudest are most likely to be heard, but the volume of the contribution does not always offer an assurance of its quality.

The purpose of this book is to offer new perspectives on AI ethics that can help illuminate the debate, and also to consider ways to progress towards solutions. Its novelty and unique contributions lie in the following:

1.   The book provides a **novel categorisation of AI** that helps to categorise technologies as well as **ethical issues**
     I propose a definition of AI in Chapter 2 that focuses on three different aspects of the term: machine learning, general AI and (apparently) autonomous digital technologies. This distinction captures what I believe to be the three main aspects of the public discussion. It furthermore helps with the next task of the book, namely the categorisation of ethical issues in Chapter 3. Based on the conceptual distinction, but also on rich empirical evidence, I propose that one can distinguish three types of ethical issues: specific issues of machine learning, general questions about living in a digital world and metaphysical issues.

2. The book proposes **human flourishing** as the basis of an **ethical framework** to deal with the ethical challenges of AI.

   The three categories of ethical issues are descriptive, which means they are derived from observations of what people perceive as ethical issues. In order to move beyond description and find a basis for practice and intervention, a normative ethical position needs to be adopted. I argue that a suitable ethical theory that can be applied to AI and provide insights that guide action is that of flourishing ethics. Flourishing ethics has three considerable advantages. First, it covers the descriptive categories of AI ethics suggested in this book. Second, it is open to other ethical theories and allows for the integration of considerations of duty, consequences and care, among others. Third, it has a distinguished history, not only in ethics broadly, but also in the ethics of computing. What flourishing ethics requires is that AI, like any other technology and tool, should contribute to human flourishing. This position is not overly contentious, provides normative guidance and is sufficiently open to be applicable to the many technologies and application domains that constitute AI ethics.

3. The book offers a **novel classification of mitigation strategies** for the ethical challenges of AI.

   Classifying ethical issues and determining a suitable ethical theory can contribute to finding possible solutions. Such solutions do not develop in a vacuum but form part of an existing discourse. I therefore review the current discussion of mitigation measures that have been proposed to deal with these issues in Chapter 4. I distinguish between several categories of mitigation options, the first referring to policy and legislation, the second to options at the organisational level and the third to guidance mechanisms for individuals.

4. The book shows that the **metaphor of an ecosystem** helps us understand the complexity of the debate and offers **insights for practical interventions.**

   Based on a rich understanding of the AI landscape, I propose the interpretation of the AI ethics debate in terms of an ecosystem. The field of AI can be pictured as a set of interlinking ecosystems which consists of many different individual actors and groups interacting in complex ways that can influence the overall system unpredictably. Returning to the idea of flourishing, I suggest asking the question: how can the AI ecosystem as a whole be shaped to foster and promote human flourishing? This interpretation of AI ethics allows actions to be prioritised and bespoke advice to be developed for individual stakeholders and stakeholder groups. Perhaps most importantly, it leads to insights into higher-level activities, namely those that are conducive to the development of the ecosystem in the desired direction of promoting human flourishing.

This novel interpretation of the AI ethics debate not only offers conceptual insights and a theoretical basis allowing us to better understand, compare and contrast various issues and options, but also provides a foundation for practical actions. These are spelled out in more detail in Chapter 5. Following an introduction to the ecosystems view of AI and its limitations, I explore its implications for possible ways of addressing ethical issues. The ecosystems view implies that interventions into the

AI ecosystem clearly delineate the boundaries of the system they apply to. Such interventions need to support the development of the ecosystem by increasing the knowledge base and capacities of its members. A final requirement for any intervention into AI ecosystems is that it needs to employ governance mechanisms that are sensitive to the non-linear and often unpredictable dynamics of the system. On this basis I then propose some activities that are likely to shape the AI ecosystem in ways that are conducive to human flourishing.

Overall, this book offers a novel perspective on the AI ethics debate. It is based on empirical insights and strong concepts that help structure the debate in a transparent and constructive manner. Very importantly, I hope that the arguments I propose point beyond AI and offer guidance that is equally applicable to whichever technology succeeds AI when the current AI hype has subsided. It thereby offers a response to Floridi's (2018) call for ways to be found of governing the digital world.

References

Condliffe J (2019) The week in tech: algorithmic bias is bad. Uncovering it is good. The New York Times. https://www.nytimes.com/2019/11/15/technology/algorithmic-ai-bias.html. Accessed 21 Sept 2020

Cossins D (2018) Discriminating algorithms: 5 times AI showed prejudice. New Sci. https://www.newscientist.com/article/2166207-discriminating-algorithms-5-times-ai-showed-prejudice/. Accessed 21 Sept 2020

The Economist (2017) Advances in AI are used to spot signs of sexuality. https://www.economist.com/science-and-technology/2017/09/09/advances-in-ai-are-used-to-spot-signs-of-sexuality. Accessed 21 Sept 2020

Floridi L (2018) Soft ethics and the governance of the digital. Philos Technol 31:1–8. https://doi.org/10.1007/s13347-018-0303-9

Isaak J, Hanna MJ (2018) User data privacy: Facebook, Cambridge Analytica, and privacy protection. Computer 51:56–59. https://doi.ieeecomputersociety.org/10.1109/MC.2018.3191268

Ledford H (2019) Millions of black people affected by racial bias in health-care algorithms. Nature 574:608–609. https://doi.org/10.1038/d41586-019-03228-6

Montreal Declaration (2018) Montréal declaration for a responsible development of artificial intelligence. Université de Montréal, Montreal. https://www.montrealdeclaration-responsibleai.com/the-declaration. Accessed 21 Sept 2020

Zuboff PS (2019) The age of surveillance capitalism: the fight for a human future at the new frontier of power. Profile Books, London

Chapter 2
Perspectives on Artificial Intelligence

Abstract A discussion of the ethics of artificial intelligence hinges on the definition of the term. In this chapter I propose three interrelated but distinct concepts of AI, which raise different types of ethical issues. The first concept of AI is that of machine learning, which is often seen as an example of "narrow" AI. The second concept is that of artificial general intelligence standing for the attempt to replicate human capabilities. Finally, I suggest that the term AI is often used to denote converging socio-technical systems. Each of these three concepts of AI has different properties and characteristics that give rise to different types of ethical concerns.

Keywords Artificial intelligence · Definitions of AI · Machine learning · Artificial general intelligence · Socio-technical systems

A good starting point for an introduction to the term "AI" is the 1956 Dartmouth summer research project on artificial intelligence, where the term was coined by McCarthy and collaborators (McCarthy et al. 2006). In their proposal for this project McCarthy et al. suggest that machines can be made to simulate "every aspect of learning or any other feature of intelligence". As features of intelligence, McCarthy et al. cite the use of language, the formation of abstractions and concepts, solving problems now reserved for humans and self-improvement.

This points to the first problem in understanding AI, namely its aim to replicate or emulate intelligence. Intelligence is itself a contested concept and it is not clear which or whose intelligence AI would have to replicate, in order to be worthy of being called AI. Biological organisms, including humans, seem to work on different principles from digital technologies (Korienek and Uzgalis 2002). Humans have access to "mental abilities, perceptions, intuition, emotions, and even spirituality" (Brooks 2002: 165). Should AI emulate all of those?

This, in turn, points to the second problem in understanding AI. Are there barriers that AI, as a digital technology, cannot overcome, aspects of intelligence that cannot be digitally replicated? This is an interesting question that has been debated for a long time (Collins 1990, Dreyfus 1992). It is ethically interesting because it has a bearing on whether AI could ever be considered an ethical subject, i.e. whether it could have

B. C. Stahl, *Artificial Intelligence for a Better Future*,
SpringerBriefs in Research and Innovation Governance,
https://doi.org/10.1007/978-3-030-69978-9_2

moral obligations in itself. This is similar to the question whether computers can think, a question that Alan Turing found "too meaningless to deserve discussion" (Turing 1950: 442) and that prompted him to propose the imitation game, also known as the Turing Test.[1]

Both problems of understanding AI – namely, what is human intelligence and which part of it might be replicable by AI – make it difficult to define AI. The conceptual subtleties of AI have led to a situation where there are many competing definitions covering various aspects (Kaplan and Haenlein 2019). The OECD (2019: 7) suggests that

> [a]n AI system is a machine-based system that can, for a given set of human-defined objectives, make predictions, recommendations, or decisions influencing real or virtual environments. AI systems are designed to operate with varying levels of autonomy.

A similarly policy-oriented definition comes from the European Commission (2020a: 2):

> AI is a collection of technologies that combine data, algorithms and computing power.

One of the most cited academic definitions is from Li and Du (2007: 1) and notes that AI combines

> a variety of intelligent behaviors and various kinds of mental labor, known as mental activities, … [to] include perception, memory, emotion, judgement, reasoning, proving, identification, understanding, communication, designing, thinking and learning, etc.

Virginia Dignum, an AI researcher who has worked extensively on ethical aspects of AI, highlights the fact that AI refers not just to artefacts, but also to an academic community. She considers

> AI to be the discipline that studies and develops computational artefacts that exhibit some facet(s) of intelligent behaviour.

> Such artefacts are often referred to as (artificial) agents. Intelligent agents are those that are capable of flexible action in order to meet their design objectives, where flexibility includes the following properties …

- Reactivity: the ability to perceive their environment, respond to changes that occur in it, and possibly learn how best to adapt to those changes;
- Pro-activeness: the ability to take the initiative in order to fulfil their own goals;
- Sociability: the ability to interact with other agents or humans.

As this book is about the ethics of AI, I propose a view of the term that is geared towards elucidating ethical concerns. Both the terms "AI" and "ethics" stand for multi-level concepts that hold a variety of overlapping but non-identical meanings. For this reason, I distinguish three aspects of the term AI, all of which have different ethical challenges associated with them.

---

[1] In the Turing Test a human participant is placed in front of a machine, not knowing whether it is operated by another human or by a computer. Can the computer's responses to the human made through the machine imitate human responses sufficiently to pass as human responses? That is what the Turing Test tries to establish.

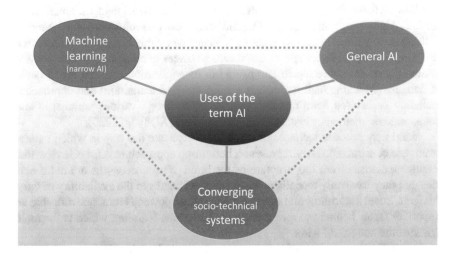

Fig. 2.1 Uses of the term "AI"

1. machine learning as the key example of a narrow understanding of AI, i.e. as a technique that successfully replicates very specific cognitive processes
2. general AI
3. AI as a synonym for converging socio-technical systems which include but go far beyond narrow AI technologies.

   Figure 2.1 gives an overview of use of the term AI that I discuss in this chapter.

2.1 Machine Learning and Narrow AI

A recent review of the AI literature by the academic publisher Elsevier (2018) suggests that there are a number of key concepts and research fields that constitute the academic discipline of AI. Based on a sample of 600 000 AI-related documents, analysed against 800 keywords, the report classified AI publications in seven clusters:

1. search and optimisation
2. fuzzy systems
3. planning and decision making
4. natural language processing and knowledge representation
5. computer vision
6. machine learning
7. probabilistic reasoning and neural networks.

   This underlines that AI is not one technology but can better be understood as a set of techniques and sub-disciplines (Gasser and Almeida 2017).

While all these clusters are recognised components of the AI field, the emphasis in current AI ethics is on machine learning and neural networks, clusters 6 and 7. Neither of these is truly novel. Machine learning has been an established part of AI research (Bishop 2006) since its inception, but recent advances in computing power and the availability of data have led to an upsurge in its application across a broad range of domains. Machine learning covers a wide array of techniques and approaches including supervised learning, Bayesian decision theory, various parametric and nonparametric methods, clustering and many others (Alpaydin 2020).

Neural networks are technologies that try to replicate the way in which natural brains are constructed. They represent a bottom-up approach to AI, i.e. a view that intelligence arises from the structure of the brain. Neural networks are not a new idea, but they have only recently achieved success thanks to the availability of large data sets, novel algorithms and increased computing power. Neural networks are an important factor behind the recent success of machine learning, which is the main driver of the current AI wave.

One particular technique of high importance is deep learning (LeCun et al. 2015), which uses different types of neural networks and has contributed to recent successes in areas such as speech recognition, visual object recognition and object detection, as well as other domains such as drug discovery and genomics (Horvitz 2017).

Machine learning, despite its impressive successes, can be characterised as an example of narrow AI. As noted earlier, this is a technique that successfully replicates very specific cognitive processes. It is not able to transfer insights easily from one domain to another. A machine learning system that has learned to distinguish cats from dogs, for example, does not automatically have the ability to recognise natural language or categorise pathology images to identify cancer. The underlying system may well be able to cover other applications but will need to be trained anew for new purposes.

For this book it is important to understand which of the characteristics that machine learning possesses are of ethical relevance. Key among them are the following:

1. *Opacity*: Machine learning algorithms and neural networks are complex to the point that their internal workings are not straightforward to understand, even for subject experts. While they remain purely technical and determined systems, it is impossible (partly because they are learning systems and therefore change) to fully understand their internal working.
2. *Unpredictability*: As a consequence of point 1, the prediction of outputs of the systems based on an understanding of the input is difficult, if not impossible.
3. *"Big data" requirements*: Machine learning systems in their current form require large training datasets and significant computer capacity to create models.

The reference to machine learning as an example of "narrow AI" suggests that there are other types of AI which are not narrow. These are typically referred to as general AI and are discussed in the next section. Before we come to these, it is important to point out that machine learning, with its use of neural networks, is not the only type of narrow AI. Other examples are decision support systems based on decision trees and fuzzy logic systems. I focus on machine learning in this book because it is the most prominent example of narrow AI right now, mostly owing to its recent successes. This is not to say that other examples of narrow AI may not gain similar prominence in future or raise other types of ethical concerns.

2.2 General AI

General AI, sometimes also referred to as "strong AI", goes back to the early days of AI research and stands for the idea that it is possible to build systems that display true human (or other higher mammal) levels of intelligence. It is also known as "good old-fashioned AI" (GOFAI). The original tenet of GOFAI was that the world could be represented through symbols and that the manipulation of these symbols would lead to intelligent behaviour (Moor and Bynum 2002). In this view the human brain was seen as a computer that performs logical operations, and the same or at least functionally equivalent ones could be performed in a digital computer (Floridi 1999).

Maybe the most interesting observation about the GOFAI project is that it has not succeeded in the 65 years since its inception. At this point there is no general AI (Babuta et al. 2020). This indicates that either its assumptions are wrong or they cannot be implemented in the type of digital computer we currently have at our disposal. There are many suggestions about why exactly GOFAI has not (yet) achieved its objectives. One suggestion is that the core of the problem is onto-logical, i.e. that the world simply cannot be represented comprehensively through symbols that are defined in a top-down manner (Smith 2019). This is the suggestion of phenomenology as expressed in an early critique of AI by Dreyfus (1972).

Another interesting question is whether the current failure of GOFAI is temporary, which would mean that we will be able to build general AI systems at some point, or whether it is fundamental, which would mean that there is some component of true intelligence that is incapable of being captured and reproduced by machines, or at least by the types of digital computers we are using today.

General AI has a strange status in this 2020 AI ethics book. On one hand it seems clear that general AI does not exist. It can therefore arguably not cause ethical concerns and can happily be ignored. On the other hand, general AI is probably the most prominent subject of discussions related to AI and ethics in science fiction, where a large number of characters represent general AI for good or ill. *2001: A Space Odyssey*, featuring the sentient computer HAL, *Blade Runner*, the Terminator movies, *I, Robot*, *WALL-E*, *Westworld* and a host of other stories are about general AI. These narratives cannot be ignored, partly because science fiction is hugely influential in guiding technical design choices, and partly because the public discussion is guided by them. High-profile interventions by celebrities and well-recognised scientists like Elon Musk and Stephen Hawking lend credence to the idea that general AI may create significant ethical risks.

In addition, general AI is of interest because many of the questions it raises are of relevance to ethics. I am agnostic about the possibility of ever creating general AI, partly because I am not sure we understand what constitutes natural intelligence and hence am not convinced that we could recognise general AI even if it appeared. The history of AI has been one of shifting goalposts, and we are now in a world where many of the early dreams of AI have been realised. Examples of successful AI implementation include the ubiquitous voice recognition that is now standard in most smart phones and the ease of organising vast amounts of data that any internet user encounters when using a search engine. Despite these successes few would say that we are anywhere near general AI. For instance, GPS systems integrated into our cars can remember our usual routes to work and suggest the most efficient one depending on current traffic conditions. They also talk to us. At the same time, we are still waiting for Lieutenant Commander Data, the android from *Star Trek: Picard*. General AI is nevertheless an important ingredient in the AI ethics debate because it brings to the fore some fundamental questions about what makes us human, and about what, if anything, the difference is between humans, other animals and artificial beings. Some of the aspects that have led to the failure of general AI so far – namely, the neglect of human nature, and of the phenomenological and existential aspects of being in the world (Heidegger 1993, Moran 1999, Beavers 2002) – are crucial for ethics and I will return to them in the next chapter.

The relevant characteristics of general AI are:

1.   Nature of intelligence: General AIGeneral AI raises the question of what constitutes intelligence.
2.   By implication, general AIGeneral AI points to fundamental questions such as:

     a.   Human nature: What does it mean to be human?
     b.   Nature of reality: What is reality?
     c.   Nature of knowledge: What can we know about reality?

General AI thus points us to some of the most fundamental philosophical questions, many of which may not have an answer or may have many inconsistent answers but are important for humans to ask to make sense of their place in the world.

While narrow AI and general AI are widely recognised concepts in the AI literature, there is another meaning of the term AI that is of high relevance to the AI ethics debate, even though it is not strictly speaking about AI in a technical sense.

2.3   AI as Converging Socio-Technical Systems

There are numerous fields of science and technology that are closely linked to AI and that are often referred to in discussions about AI. Some of these are technologies that produce the data that machine learning requires, such as the internet of things. Others are technologies that can help AI to have an effect on the world, such as robotics (European Commission 2020b). One could also use the term "smart information system" (SIS) to denote this combination of several types of technologies, which typically are based on machine learning and big data analytics (Stahl and Wright

2018). In practice AI rarely appears as a stand-alone technology but is usually linked to and embedded in other technologies.

The distinction between different technologies is increasingly difficult. Fifty years ago, a computer would have been a readily identifiable large machine with clearly defined inputs, outputs and purposes. Since then the increasing miniaturisation of computing devices, the introduction of mobile devices, their linking through networks and their integration into communications technologies have led to a situation where computing is integrated into most technical devices and processes. AI tends to form part of these technical networks.

Some authors have used the abbreviation NBIC (nano, bio, information and cognitive technologies) to denote the apparent convergence of these seemingly different technologies (Khushf 2004, Van Est et al. 2014). AI and brain-related technologies have a central role in this convergence.

Perhaps not surprisingly, there is much work that links AI with neuroscience, the scientific study of the brain and the nervous system. Since the brain is the seat of human intelligence, research on the brain is likely to be relevant to understanding artificial as well as natural intelligence. AI has always drawn from our understanding of the brain, with artificial neural networks being a prominent example of how neuroscientific insights have influenced AI development. At present there is much interest in what neuroscience and machine learning can learn from each other (Marblestone et al. 2016, Glaser et al. 2019) and how neuroscience and AI research, in their further progress, can support each other (Hassabis et al. 2017). One hope is that neuroscientific insights may help us move beyond narrow AI to general AI, to the development of machines that "learn and think like people" (Lake et al. 2017).

The term "AI" in this context is thus used as shorthand for technical systems and developments that have the potential to grow together, to support and strengthen one another. Crucially, these systems are not just technical systems but *socio*-technical systems. While this is true for any technical system (they never come out of nothing and are always used by people) (Mumford 2006), it is particularly pertinent for the converging technologies that include AI. Examples of such socio-technical systems include most of the high-profile examples of AI, such as autonomous vehicles, embedded pattern recognition – for example, for the scrutiny of CVs for employment purposes – and predictive policing. All of these have a narrow AI at their core. What makes them interesting and ethically relevant is not so much the functioning of the AI, but the way in which the overall socio-technical system interacts with other parts of social reality.

This use of the term "AI" to denote socio-technical systems containing AI and other technologies points to some characteristics of these technologies that are ethically relevant. These socio-technical systems appear to be autonomous, i.e. they create outputs that affect people in ways that do not allow responsibility to be ascribed to human beings. This does not imply a strong concept of the autonomy of AI, a concept I will return to in the following chapter, but rather a lack of visible oversight and control. For instance, if embedded pattern recognition is used to scan CVs to identify candidates suitable for interviewing, the system is not an example of strong autonomy (as a human short-lister would be), but the ethical issues in terms of oversight are still obvious.

Another important aspect of these systems is that they structure the space of options that individuals have. Coeckelbergh (2019) uses the metaphor of theatre roles. Drawing on Goffman (1990), Coeckelbergh argues that human actions can be seen as embodied performances. The scope of content of these performances is structured by what is available on the stage. AI-driven socio-technical systems take the role of the theatre, often of the director. Even if they do not directly instruct humans as to what they should do (which is also often the case; think of the Uber driver receiving her instructions from her phone), they determine what can or cannot be done. Where humans are not aware of this, such a structuring of options can be seen as a covert manipulation of human actions. And, given the economic and social reach and importance of these technologies, the social impact of these systems can be significant. For instance, the use of an internet search engine and the algorithms used to determine which findings are displayed structure to a large extent what users of this search engine are aware of with regard to the search. Similarly, the information made available to social media users, typically prioritised by AI, can strongly influence people's perception of their environment and thereby promote or limit the prevalence of conspiracy theories. To summarise, the AI-enabled socio-technical systems have the following characteristics.

1. *Autonomy*: AI socio-technical systems lead to consequences for humans that are not simply results of identifiable actions of human beings.
2. *Manipulation*: AI socio-technical systems structure human options and possible actions, often in ways that humans do not realise.
3. *Social impact*: Consequences for individuals and society of the use of AI socio-technical systems can be significant.

Figure 2.2 provides a graphical representation of the features of the different meanings of AI discussed in this chapter.

This view of AI and its sub-categories helps us better understand and deal with the ethical issues currently discussed in the context of AI. It should be clear, however, that I do not claim that it is the only way of categorising AI, nor would I argue that the three categories are distinctly separate. Machine learning may well hold the key to general AI and it certainly forms part of the converging socio-technical systems. Should general AI ever materialise, it will no doubt form a part of new socio-technical systems. The purpose of the distinction of the three aspects is to show that there are different views of AI that point to different characteristics of the term, which, in turn, raises different ethical issues. It therefore facilitates engagement with ethical issues.

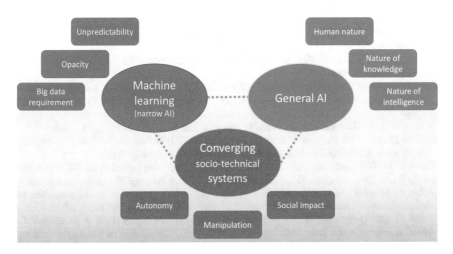

Fig. 2.2 Key characteristics of the different uses of the term "AI"

References

Alpaydin E (2020) Introduction to machine learning. The MIT Press, Cambridge MA

Babuta A, Oswald M, Janjeva A (2020) Artificial intelligence and UK national security: policy considerations. RUSI Occasional Paper. Royal United Services Institute for Defence and Security Studies, London. https://rusi.org/sites/default/files/ai_national_security_final_web_version.pdf. Accessed 21 Sept 2020

Beavers AF (2002) Phenomenology and artificial intelligence. Metaphilosophy 33:70. https://doi.org/10.1111/1467-9973.00217

Bishop CM (2006) Pattern recognition and machine learning. Springer Science+Business Media, New York

Brooks RA (2002) Flesh and machines: how robots will change us. Pantheon Books, New York

Coeckelbergh M (2019) Technology, narrative and performance in the social theatre. In: Kreps D (ed) Understanding digital events: Bergson, Whitehead, and the experience of the digital, 1st edn. Routledge, New York, pp 13–27

Collins HM (1990) Artificial experts: social knowledge and intelligent systems. MIT Press, Cambridge MA

Dreyfus HL (1972) What computers can't do: a critique of artificial reason. Harper & Row, New York

Dreyfus HL (1992) What computers still can't do: a critique of artificial reason, revised edn. MIT Press, Cambridge MA

Elsevier (2018) Artificial intelligence: how knowledge is created, transferred, and used. Trends in China, Europe, and the United States. Elsevier, Amsterdam. https://www.elsevier.com/__data/assets/pdf_file/0011/906779/ACAD-RL-AS-RE-ai-report-WEB.pdf. Accessed 22 Sept 2020

European Commission (2020a) White paper on artificial intelligence: a European approach to excellence and trust. European Commission, Brussels. https://ec.europa.eu/info/sites/info/files/commission-white-paper-artificial-intelligence-feb2020_en.pdf. Accessed 22 Sept 2020

European Commission (2020b) Report on the safety and liability implications of artificial intelligence, the internet of things and robotics. European Commission,

Brussels. https://ec.europa.eu/info/files/commission-report-safety-and-liability-implications-ai-internet-things-and-robotics_en. Accessed 22 Sept 2020

Floridi L (1999) Information ethics: on the philosophical foundation of computer ethics. Ethics Inf Technol 1:33–52

Gasser U, Almeida VAF (2017) A layered model for AI governance. IEEE Internet Comput 21:58–62. https://doi.org/10.1109/MIC.2017.4180835

Glaser JI, Benjamin AS, Farhoodi R, Kording KP (2019) The roles of supervised machine learning in systems neuroscience. Prog Neurobiol 175:126–137. https://doi.org/10.1016/j.pneurobio.2019.01.008

Goffman E (1990) The presentation of self in everyday life, New Ed edn. Penguin, London

Hassabis D, Kumaran D, Summerfield C, Botvinick M (2017) Neuroscience-inspired artificial intelligence. Neuron 95:245–258. https://doi.org/10.1016/j.neuron.2017.06.011

Heidegger M (1993) Sein und Zeit, 14th edn. Max Niemeyer Verlag GmbH & Co KG, Tübingen

Horvitz E (2017) AI, people, and society. Science 357:7. https://doi.org/10.1126/science.aao2466

Kaplan A, Haenlein M (2019) Siri, Siri, in my hand: who's the fairest in the land? On the interpretations, illustrations, and implications of artificial intelligence. Bus Horiz 62:15–25

Khushf G (2004) Systems theory and the ethics of human enhancement: a framework for NBIC convergence. In: Roco MC, Montemagno CD (eds) Coevolution of human potential and converging technologies. New York Academy of Sciences, New York, pp 124–149

Korienek G, Uzgalis W (2002) Adaptable robots. Metaphilosophy 33:83–97

Lake BM, Ullman TD, Tenenbaum JB, Gershman SJ (2017) Building machines that learn and think like people. Behav Brain Sci 40:e253. https://doi.org/10.1017/S0140525X16001837

LeCun Y, Bengio Y, Hinton G (2015) Deep learning. Nature 521:436–444

Li D, Du Y (2007) Artificial intelligence with uncertainty. Chapman and Hall/CRC, Boca Raton FL

Marblestone AH, Wayne G, Kording KP (2016) Toward an integration of deep learning and neuroscience. Front Comput Neurosci 10:94. https://doi.org/10.3389/fncom.2016.00094

McCarthy J, Minsky ML, Rochester N, Shannon CE (2006) A proposal for the Dartmouth summer research project on artificial intelligence. AI Magazine 27:12–14. https://doi.org/10.1609/aimag.v27i4.1904

Moor JH, Bynum TW (2002) Introduction to cyberphilosophy. Metaphilosophy 33:4–10

Moran D (1999) Introduction to phenomenology, 1st edn. Routledge, London

Mumford E (2006) The story of socio-technical design: reflections on its successes, failures and potential. Inf Syst J 16:317–342. https://doi.org/10.1111/j.1365-2575.2006.00221.x

OECD (2019) Recommendation of the council on artificial intelligence. OECD/LEGAL/0449

Smith BC (2019) The promise of artificial intelligence: reckoning and judgment. The MIT Press, Cambridge MA

Stahl BC, Wright D (2018) Ethics and privacy in AI and big data: implementing responsible research and innovation. IEEE Secur Priv 16:26–33. https://doi.org/10.1109/MSP.2018.2701164

Turing AM (1950) Computing machinery and intelligence. Mind 59:433–460

Van Est R, Stemerding D, Rerimassie V, Schuijff M, Timmer J, Brom F (2014) From bio to NBIC: from medical practice to daily life. Rathenau Instituut, The Hague

Chapter 3
Concepts of Ethics and Their Application to AI

Abstract Any discussion of the ethics of AI needs to be based on a sound understanding of the concept of ethics. This chapter therefore provides a brief overview of some of the key approaches to ethics with a particular emphasis on virtue ethics and the idea of human flourishing. The chapter reviews the purposes for which AI can be used, as these have a bearing on an ethical evaluation. Three main purposes are distinguished: AI for efficiency, optimisation and profit maximisation, AI for social control and AI for human flourishing. Given the focus on human flourishing in this book, several theoretical positions are introduced that provide insights into different aspects and ways of promoting human flourishing. The chapter concludes with a discussion of the currently widespread principle-based approach to AI ethics.

Keywords Ethical theory · Human flourishing · Purposes of AI · Ethical principles for AI

Ethical issues of AI are hotly debated and sometimes contested. In order to understand what they are and why they might be considered ethical issues, and to start thinking about what can or should be done about them, I start with an introduction to ethics, which is then followed by an empirically based discussion of current ethical issues of AI.

At its most basic level, ethics has to do with good and bad, with right and wrong. However, the term "ethics" is much more complex than that and the same word is used to cover very different aspects of the question of right and wrong. Elsewhere (Stahl 2012), I have proposed the distinction of four different levels, all of which are covered by the term "ethics":

1. Moral intuition, expressed in a statement of the sort: "This is right," or "This is wrong."
2. Explicit morality, expressed in general statements like "One should always /never do this."
3. Ethical theory, i.e. the justification of morality drawing on moral philosophy expressed in statements like "Doing this is right/wrong because ..."
4. Metaethics, i.e. higher-level theorising about ethical theories.

© The Author(s) 2021
B. C. Stahl, *Artificial Intelligence for a Better Future*,
SpringerBriefs in Research and Innovation Governance,
https://doi.org/10.1007/978-3-030-69978-9_3

This view of ethics is compatible with other views, notably the frequently suggested distinction between applied ethics, normative ethics and metaethics. It also accommodates the typical introduction to ethics that one can find in technology ethics textbooks (Van de Poel and Royakkers 2011), notably the dominant ethical theories of deontology and consequentialism.

3.1 Ethical Theories

Ethical theories are attempts to find an answer to the question: what makes an action ethically better or worse than an alternative action? Prominent examples of ethical theories include consequentialism and deontology. (I shall return to virtue ethics later.) Both of these originated during the Enlightenment period (mainly in the 18th century). They aim to provide clear rules that allow us to determine the ethical quality of an action. Consequentialist theories focus on the *outcomes* of the action for this evaluation. The various approaches to utilitarianism going back to Jeremy Bentham (1789) and John Stuart Mill (1861) are the most prominent examples. They are based on the idea that one can, at least in theory, add up the aggregate utility and disutility resulting from a particular course of action. The option with the highest net utility, i.e. utility minus disutility, is the ethically optimal one.

Deontology, on the other hand, is based on the principle that the basis of the ethical evaluation of an action is the duty of the agent executing it. The most prominent representative of this position is Immanuel Kant (1788, 1797), who formulated the so-called categorical imperative. The most often quoted formulation of the categorical imperative says "Act only on that maxim by which you can at the same time will that it should become a universal law" (translation, quoted in Bowie 1999: 14). This categorical imperative stops agents from rationalising exemptions for themselves. The interesting aspect of such a position for our purposes is that this view of ethics pays no immediate attention to the consequences of an action, but exclusively focuses on the motivation for undertaking it.

It is important to underline, however, that deontology and utilitarianism are not the only ethical theories that can be applied to AI, and to technology more broadly. In addition to virtue ethics, to which I will return shortly, there are other general ethical approaches such as the feminist ethics of care (Gilligan 1990) and ethics based on various religions. Applying ethical theories to particular application areas has resulted in rich discourses of concepts such as computer ethics (Bynum and Rogerson 2003, Bynum 2008a, van den Hoven 2010), information ethics (Capurro 2006, Floridi 2006) and technology ethics (Brey 2011) that are relevant to AI.

Entire libraries have been written about philosophical ethics, and I cannot hope to do justice to the many and rich nuances of ethical thinking. It may nevertheless be helpful to outline how ethics links to the human condition. This can explain some of the characteristics of ethics and it can shed light on whether or to what degree non-human artificial agents can be ethical subjects.

A key to understanding ethics, I believe, is that humans recognise that we all, despite many and far-reaching differences, have much in common. We could call this state "the shared features of the human condition". Human beings are fundamentally social. Without social structures and support we would not only die as infants, but also fail to develop the language and thus the conceptual understanding of the world around us that allow us to live our lives. We are possibly the only species that not only recognises that we exist but also knows that we are fundamentally vulnerable and mortal. We not only know this, but we feel it in profound ways, and we recognise that we share these feelings with other humans. The shared fate of certain death allows us to see the other as someone who, no matter how different from us they are, has some basic commonalities with us. We have empathy with others based on our experiences and the assumptions that they are like us. And just as we share the knowledge of death, we also share the experience of hope, of joy, of the ability to (more or less) freely develop projects and shape our world. This world is not just a physical world, but predominantly a social one, which is constructed using the unique capabilities of human language. Ethics is then a way to shape an important part of this social world in ways that take into account the shared aspects of human nature.

This description of human nature and the human condition has direct implications for the concept of ethics and what can count as "being ethical". Ethics does not exclusively reside in an action or an intention. Ethics is part of *being* in the world, to use a Heideggerian term (Heidegger 1993). It is characterised by an agent's ability not only to perceive different possible states of the world and decide between conceivable options, but to do so with a view to the meaning of such a decision for her own world and also for the world at large. This implies that the agent is consciously situated in this world, and understands it, but also has an emotional relationship to it and the fellow agents who co-constitute this world. Such an agent may very well make use of deontological or utilitarian ethical theories, but she does so in a reflective way as an agent who has a commitment to the world where these principles are applied.

This brief introduction to my ethical position points to the idea of human flourishing, which will become vital in later parts of this book: human flourishing linked to *being* in the world, understanding the limits of the human condition and the essential socialness of humans, which requires empathy. Of course, I realise that there are people who have no or little empathy, that abilities to interact socially and use language differ greatly, that many of these aspects apply to some degree also to some animals. Yet, to substantiate my position in AI ethics and the main ideas of this book, it is important that I do not draw inordinately on deontology and utilitarianism, but rather take into account a wider range of sources, and in particular virtue ethics.

3.2 AI for Human Flourishing

Current approaches to philosophical ethics as represented by consequentialism and deontology are largely rational and theoretical endeavours and mostly at home in academic philosophy departments. Ethics, however, has traditionally had a much broader meaning. For the ancient Greeks, philosophy was not just an intellectual endeavour but an attempt to find ways to live the "good life", the answer to the question: how should I live (Annas 1993)? The major philosophical schools of ancient Greece agreed that the cosmos had a purpose and that the individual good life, resulting in happiness (Aristotle 2007), was predicated on people fulfilling their role in society. This is the basis of virtue ethics, which is most prominently associated with Aristotle (2007) but whose main tenets are widely shared across philosophical schools. The focus of this approach to ethics is not so much the evaluation of the anticipated outcomes of an individual act or their intention, but providing guidance for the individual to help them develop a virtuous character.

I do not want to overly romanticise ancient Greece, whose acceptance of slavery and misogyny are not acceptable. However, virtue ethics as an approach to ethics has significant appeal, probably because it offers to provide guidance not only on individual problems but on how we should live our lives. This may explain why it has returned to prominence since the end of the 20th century and seen attempts to translate it into modern contexts (MacIntyre 2007).

Terry Bynum is one of several scholars who have succeeded in translating the ancient principles of virtue ethics into a modern technology-saturated context. He suggests the development of a "flourishing ethics" (Bynum 2006) which draws from Aristotelian roots. Its key tenets are:

1. Human flourishing is central to ethics.
2. Humans as social animals can only flourish in society.
3. Flourishing requires humans to do what we are especially equipped to do.
4. We need to acquire genuine knowledge via theoretical reasoning and then act autonomously and justly via practical reasoning in order to flourish.
5. The key to excellent practical reasoning and hence to being ethical is the ability to deliberate about one's goals and choose a wise course of action.

Bynum (2008b) has shown that these principles of virtue ethics are relevant to and have informed ethical considerations of information technology since its early days and can be found in the work of Norbert Wiener (1954), one of the fathers of digital technology.

Much research has been undertaken to explore how principles of virtue ethics can be applied to technology and how we can live a virtuous life in a technologically driven society. An outstanding discussion of virtue ethics in the context of digital technologies is provided by Vallor (2016), and, given that my approach relies heavily on her discussion, I will return to it later with reference to human flourishing.

As Bynum points out, people are endowed with different skills and strengths. Flourishing includes excellence in pursuit of one's goals, which implies that there are as many ways of flourishing as there are combinations of skills. Flourishing is thus not a one-size-fits-all concept but needs to be filled with life on an individual level. Before I return to a more detailed discussion of the concept of flourishing, I now want to discuss the motivations behind and purposes of developing, deploying and using AI, as these have a direct bearing on the ethical evaluation of AI socio-technical systems.

3.3 Purposes of AI

Understanding the purpose and intention of AI is important when thinking about the ethics of AI. Digital technologies, as pointed out earlier, are highly flexible and open to interpretation. They are logically malleable. They can thus be used for an infinity of purposes, which may or may not be aligned with the intention of the original developers and designers. Despite this openness of AI, it is still possible to distinguish different purposes that determine the design, development and use of systems. I distinguish three main purposes: AI for efficiency, AI for social control and lastly, as an alternative and complement to the two initial ones, AI for human flourishing (see Fig. 3.1).

When looking at current policy documents covering AI, one typically finds a mixture of all three of these motivations: AI can *improve efficiency*, which will lead to cost savings and thereby to economic benefits, which will trickle down, and people's lives will get better. A report to the President of the United States set the tone by highlighting the economic advantages and suggesting that "AI has the potential to double annual economic growth rates in the countries analyzed by 2035" (Executive Office of the President 2016). The European Commission expects that "AI could spread across many jobs and industrial sectors, boosting productivity, and yielding strong positive growth" (Craglia et al. 2018). And a committee of the United Kingdom's House of Lords hopes that "AI could spread across many jobs and industrial sectors, boosting productivity, and yielding strong positive growth" (House of Lords 2018).

A very different view of the use of technology including AI is to see it as a way of exerting *social control*. Rapidly growing abilities to collect data, in conjunction

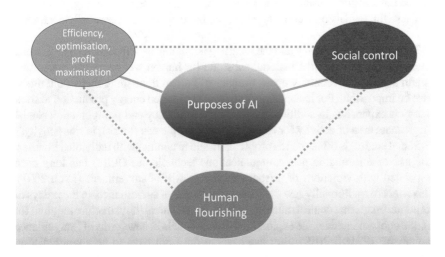

Fig. 3.1 Possible purposes of AI

with AI's ability to detect patterns and correlations between variables, allow for new ways of controlling human behaviour. This can be done in subtle ways, using the idea of "nudging" based on behavioural economics (Mullainathan and Thaler 2000, Camerer et al. 2004) or it can be done more vigorously, as for example in the Chinese social credit scoring system (Creemers 2018, Liu 2019).

> The system intends to monitor, rate and regulate the financial, social, moral and, possibly, political behavior of China's citizens – and also the country's companies – via a system of punishments and rewards. The stated aim is to "provide the trustworthy with benefits and discipline the untrustworthy." (Bartsch and Gottske nd)

AI as social control can also breach the limits of legality, as happened in the Facebook–Cambridge Analytica case, where social media data was used to illegitimately influence the outcome of democratic elections (Isaak and Hanna 2018). Zuboff (2019) offers a forceful argument that social control is a driving force and a necessary condition of success of what she calls "surveillance capitalism". In her analysis she does not focus on the term AI, but her description of the way in which new business models have developed and facilitated enormous profits is fully aligned with the concept of AI as converging socio-technical systems (see Fig. 3.1).

The third purpose of using AI, drawing on the earlier discussion of ethics, is to employ it for *human flourishing*. This means that AI is developed and deployed in ways that promote human flourishing. It can be used as a tool that helps individuals and groups identify worthwhile goals and supports them in their pursuit of excellence in achieving these goals. There are a number of suggestions on how to ensure that AI has positive consequences for individuals and societies, which is part of this third purpose of using AI for human flourishing: for example, attempts to construct a "good AI society" (Cath et al. 2016) or the discourse on AI for good that I discuss in more detail below in the section on the benefits of AI.

The three different views of the purpose of AI are represented in Fig. 3.1.

These three goals may come into conflict, but they are not necessarily contradictory.

The pursuit of efficiency and the resulting economic benefits can lead to a strong economy that provides the material substrate for human wellbeing. By generating wealth an efficient economy opens avenues of human flourishing that would otherwise be impossible. For instance, a move from coal-based energy production to solar energy is expensive. In addition, the pursuit of efficiency and profit creation can be a legitimate area of activity for excellence, and people can flourish in this activity.

Social control is often seen as problematic and in conflict with individual liberties. The use of information and communications technologies (ICTs) has long been associated with violations of privacy and the growth of surveillance (Lyon 2001). This concern traditionally saw the state as the source of surveillance. In these days of corporate giants that control much of the data and technical infrastructure required for AI, the concern includes the exploitation of individuals in new forms of "surveillance capitalism" (Zuboff 2019). But, again, there does not have to be a contradiction between social control and human flourishing. Humans as social beings need to define ways of collaborating, which includes agreement on moral codes, and these

Fig. 3.2 Overlap of
purposes of AI

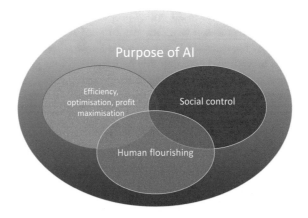

need to be controlled and enforced in some way. While nudging as a policy instrument is contentious, it can be and often is used to promote behaviours that are conducive to flourishing, such as promoting a healthier lifestyle. Used especially in the United Kingdom, Australia, Germany and the US (Benartzi et al. 2017), nudging involves government-led campaigns to achieve given targets, for instance higher vaccination rates. For example, a US campaign involved sending out planning prompts for flu vaccination to citizens, which increased vaccination rates by 4.2% (ibid).

In the technology domain AI can be used to promote privacy awareness (Acquisti 2009), arguably a condition of flourishing. As I write these sentences, much of the world is under lockdown due to the COVID-19 pandemic. In the UK there is a heated debate around apps to be used to support the tracking and tracing of infected individuals (Klar and Lanzerath 2020). What this shows is that even forced social control through digital technologies may in some circumstances be conducive to human flourishing, for example, if it can help save lives and allow society to function. A Venn-type diagram may therefore be a better representation of the relationship of the three purposes (Fig. 3.2).

I must emphasise that the three purposes of AI listed in Figures 3.1 and 3.2 are not intrinsically contradictory, but rather describe the main fields of emphasis or different directions of travel that can guide the development and deployment of AI. My proposal is that the explicit aim to do the ethically right thing with AI can be described with reference to human flourishing.

This is not a novel insight. It draws from the ancient Greek philosophers and has been applied to ICT for decades. It has also been applied to AI. Virginia Dignum (2019: 119), for example, states: "Responsible Artificial Intelligence is about human responsibility for the development of intelligent systems along fundamental human principles and values, to ensure human flourishing and well-being in a sustainable world." Mark Coeckelbergh (2019: 33) voices a similar view when he states that we "need a positive and constructive ethics of AI, which is not only about regulation in the sense of constraints but which also concerns the question of the good life and human and societal flourishing". The principle of this argument is unproblematic and can also be found in AI policy proposals (ALLEA and Royal Society 2019). Who, after all, would say that they want to use AI to limit human flourishing? However, it raises the questions: how can we know whether human flourishing is promoted or achieved, and how can this be translated into practice? In order to answer these questions, I will now look at some theoretical positions on technology and its role in the world.

3.4 Theoretical Perspectives on Human Flourishing

Flourishing ethics is part of the tradition of virtue ethics and its historical roots in Aristotelian ethics. In order to answer the question, "How can we understand flourishing in practical terms?" it is helpful to look at other positions that share the aim of promoting human flourishing. Three positions that have been applied to technology, or that were developed specifically with research and technology development in mind, are important in this context: critical theory of technology, capability theory and responsible research and innovation. Each of these three offers an established theoretical approach that is consistent with human flourishing, and each has led to a wealth of insights into how flourishing can be observed and promoted..

Critical theory of technology is my first example of a theoretical approach relevant to AI that encompasses flourishing. Critical theory has a number of different possible roots. In its European spirit it tends to trace its origins to Marx's criticism of capitalism. There is a recurrence of Marxist thinking in relation to digital technologies (Greenhill and Wilson 2006, Fuchs and Mosco 2017). However, much of critical theory of technology uses later developments of critical theory, notably of the Frankfurt School (Wiggershaus 1995). Andrew Feenberg's (1993, 1999) work is probably the best-known example of the use of critical theory to understand modern technology. In addition, there has been a long-standing discussion of critical theory in the field of information systems, which draws on further theoretical traditions, such as postcolonialism (Mayasandra et al. 2006) and postmodernism (Calás and Smircich 1999).

Elsewhere I have argued that one central combining feature of the various different views of critical theory is that they aim to promote emancipation (Stahl 2008). The emancipatory intention of critical research, i.e. research undertaken in the critical tradition, means that resulting research cannot be confined to description only, but attempts to intervene and practically promote emancipation (Cecez-Kecmanovic 2011). Myers and Klein (2011), drawing on Alvesson and Willmott (1992), see emancipation as facilitating the realisation of human needs and potential, critical self-reflection and associated self-transformation. The concept of emancipation seems very close to the principle of human flourishing discussed earlier. My reason for bringing critical theory into this discussion is that critical theory has developed a set of tools and a high degree of sensitivity for understanding factors that can impede emancipation. Because of its roots in Marxist ideology critique, critical theory is well positioned to point to the factors limiting emancipation and flourishing that arise from the current socio-economic system, from labour processes and from capitalist modes of production. As will be seen later, these constitute probably the largest set of ethical issues associated with AI.

A second theoretical position worth highlighting in the context of human flourishing is *capability theory*. Capability theory has roots in philosophy and economics and is strongly associated with Amartya Sen (2009) and Martha Nussbaum (2011). The capability approach originated in development economics and the desire to find better ways of describing human development than purely financial and aggregate measures such as the gross domestic product. It is also directly linked to and based on the Aristotelian notion of flourishing (Johnstone 2007), and thus immediately relevant to a discussion of the ethics of AI and human flourishing.

The reason for highlighting the capability approach is that it has a history of application to information technologies (Oosterlaken and van den Hoven 2012), often in the context of studies of ICT for development and its focus on marginalised and vulnerable populations (Kleine 2010). It can thus be used as a way of sharpening the focus on the impact that AI can have on such populations. In addition, the communities working with the capability approach have developed tools for improving human functioning and freedoms and for measuring outcomes that have been recognised at a political level, notably by the United Nations. It is therefore suited to the creation of metrics that can be used to assess whether AI applications and uses benefit human flourishing.

The final theoretical position relevant to AI ethics and human flourishing is that of *responsible research and innovation* (RRI). RRI is a concept that has gained prominence in research and innovation governance since around the early 2010s. It has been defined as the "on-going process of aligning research and innovation to the values, needs and expectations of society" (European Union 2014). There are different interpretations of RRI (Owen and Pansera 2019), including that of the European Commission (2013), which consists of six pillars or keys (engagement, gender equality, science education, ethics, open access and governance), and that of the UK's Engineering and Physical Sciences Research Council (Owen 2014), represented by the AREA acronym (anticipate, reflect, engage and act), which is based on Stilgoe et al. (2013).

A much-cited definition of RRI proposed by Von Schomberg (2013: 63) sees RRI as

> a transparent, interactive process by which societal actors and innovators become mutually responsive to each other with a view to the (ethical) acceptability, sustainability and societal desirability of the innovation process and its marketable products (in order to allow a proper embedding of scientific and technological advances in our society).

The reference to RRI is helpful in the context of AI ethics because it puts research and innovation explicitly into the societal context. The idea that the process and product of research and innovation should be acceptable, sustainable and societally desirable can be read as implying that they should be conducive to human flourishing. RRI can thus be understood as a way of promoting and implementing human flourishing. RRI is important in the context of this book because it is established as a term in research funding and familiar to policymakers. A recent proposal by the European Parliament puts heavy emphasis on RRI as a way to ensure ethical sensitivity in future AI research, development and deployment. The European Parliament (2020: 6) suggests that "the potential of artificial intelligence, robotics and related technologies … should be maximized and explored through responsible research and innovation".

Human flourishing in the broad sense used here is something that I believe most people can sign up to. It does not commit us to a particular way of life or require the adoption of a particular ethical position. It does not prevent us from using other ethical theories, including deontology and utilitarianism, to assess ethical questions (Bynum 2006). It is compatible with various theoretical positions beyond the three (critical theory, capability theory, RRI) introduced here. The choice of human flourishing was guided by the need to find an ethical language that can find traction across disciplinary, national, cultural and other boundaries. AI technologies are global and pervasive, but they have an impact at the local and individual level. An approach to the ethics of AI that aims to provide general guidance therefore needs to be able to build bridges across these many global divides, which I hope the idea of flourishing does.

3.5   Ethical Principles of AI

The main thesis of this book is that flourishing ethics can enlighten AI ethics and provide guidance in the development of practical interventions. The majority of currently existing guidelines were not drafted from one theoretical viewpoint but tend to use a set of ethical principles or values. What are these values?

The most comprehensive review of AI ethics guidelines published so far (Jobin et al. 2019) lists the following ethical principles: transparency, justice and fairness, non-maleficence, responsibility, privacy, beneficence, freedom and autonomy, trust, sustainability, dignity and solidarity. Each of these is comprised of components. Transparency, for example, refers to related concepts such as explainability, explicability, understandability, interpretability, communication and disclosure. The relationship between these concepts is not normally well defined and they can refer to different ethical positions. Elsewhere we have tried to clarify their normative implications (Ryan and Stahl 2020).

Another example, the ethics guidelines for trustworthy AI proposed by the EU's High-Level Expert Group on Artificial Intelligence (2019), has a tiered level of principles. The expert group proposes a framework for trustworthy AI that consists of lawful AI (which they do not cover), ethical AI and robust AI. This framework is based on four ethical principles: respect for human autonomy, prevention of harm, fairness and explicability. From these principles they deduce seven key requirements for the realisation of trustworthy AI, namely:

1. human agency and oversight
2. technical robustness and safety
3. privacy and data governance
4. transparency
5. diversity, non-discrimination and fairness
6. social and environmental wellbeing
7. accountability.

From these they then develop assessment methods for trustworthy AI and policy recommendations.

It is easy to see the attraction of this principle-based approach. It avoids making strong commitments to typically contested ethical theories. The principles themselves are generally uncontroversial, thereby offering the opportunity of a consensus. Maybe most importantly, the principle-based approach has been the basis of biomedical ethics, the field of ethics with the longest history of high-visibility public debate and need for societal and political intervention. Biomedical ethics in its modern form resulted from the Nazi atrocities committed during research on humans in concentration camps and the Nuremberg Code (Freyhofer 2004) that paved the way for the Declaration of Helsinki (World Medical Association 2008). It was codified and operationalised through the Belmont Report (National Commission for the Protection of Human Subjects of Biomedical and Behavioral Research 1979), which established the principles of biomedical ethics that remain dominant in the field (Beauchamp and Childress 2009): autonomy, justice, beneficence and non-maleficence.

Biomedical ethics has been hugely influential and underpins discussion of the human rights of patients (Council of Europe 1997). One crucial aspect of biomedical ethics is that it has been implemented via the well-established process of research ethics, based on ethics review, conducted by institutional review boards or research ethics committees, overseen by regional or national boards and strongly sanctioned by research funders, publishers and others.

There can be little doubt that this institutional strength of biomedical (research) ethics is a central factor guiding the AI ethics debate and leading to a principle-based approach that can be observed in most guidelines. This dominant position nevertheless has disadvantages. Biomedical ethics has been criticised from within the biomedical field as being overzealous and detrimental to research (Klitzman 2015). Empirical research on biomedical research ethics has shown inconsistency with regard to the application of principles (Stark 2011). And while largely uncontested in the biomedical domain, though not completely (Clouser and Gert 1990), the applicability of this approach to ethics in other domains, such as the social sciences, has been vehemently disputed (Schrag 2010).

There are two aspects from this discussion worth picking up for AI ethics. Firstly, there is the question of the implicit assumptions of biomedical ethics and their applicability to AI. Biomedical ethics was developed primarily to protect the rights of patients and research participants. This is no doubt transferable to AI, where the individuals on the receiving end of AI systems are worthy of protection. But because biomedical research predominantly aims to understand diseases with a view to finding cures, biomedical ethics is much less concerned with the purpose of the research. It is usually taken for granted that biomedical research pursues an ethically commendable goal: that of contributing to human health and thus to human wellbeing. Ethical concerns therefore do not arise from this goal itself but only from ways of achieving it. In the case of technical research, including AI research, it is not at all obvious that this implicit premise of biomedical research is applicable. The assumption that the research itself and its intended consequences are ethically acceptable and desirable is in need of much more questioning and debate, casting doubt on whether the process-oriented and principle-based biomedical research ethics process is a suitable one to base AI ethics on.

Secondly, biomedical principlism (Beauchamp and Childress 2009) leaves open the question of how to deal with conflicts between principles. This is a well-established problem for any ethical approach that is based on a set of non-hierarchical principles or values. In most cases it is possible to imagine situations where these come into conflict. Looking at the principles used in AI, it is, for example, easy to imagine a case where the principle of transparency would come into conflict with the principle of privacy. In order to successfully guide action or decision, the approach therefore needs to find ways of dealing with such conflicts. In addition, principlism has been criticised for being overly close to its US origins and not generalisable across the world (Schroeder et al. 2019).

Framing AI ethics in terms of human flourishing can address both concerns. By offering an overarching ethical ambition it proposes a point of comparison that can help address value conflicts. It also aligns more closely to 21st-century research ethics, which has been moving away from Western principles to global values (Schroeder et al. 2019). And it furthermore offers a perspective that does not take for granted that all research and technology innovation is desirable per se, but clearly posits flourishing as the overarching goal.

References

Acquisti A (2009) Nudging privacy: the behavioral economics of personal information. IEEE Secur Priv 7:82–85. https://doi.org/10.1109/MSP.2009.163

ALLEA, Royal Society (2019) Flourishing in a data-enabled society. https://royalsociety.org/-/media/policy/Publications/2019/28-06-19-flourishing-in-data-enabled-society.pdf?la=en-GB&hash=D521F71EB21F9369FAC26D7E1313398A. Accessed 23 Sept 2020

Alvesson M, Willmott H (1992) On the idea of emancipation in management and organization studies. Acad Manage Rev 17:432–464

Annas J (1993) The morality of happiness, New edn. Oxford University Press, New York

Aristotle (2007) The Nicomachean ethics. Filiquarian Publishing, Minneapolis

Bartsch B, Gottske M (nd) China's social credit system. Bertelsmann Stiftung. https://www.bertelsmann-stiftung.de/fileadmin/files/aam/Asia-Book_A_03_China_Social_Credit_System.pdf. Accessed 25 Sept 2020

Beauchamp TL, Childress JF (2009) Principles of biomedical ethics, 6th edn. Oxford University Press, New York

Benartzi S, Beshears J, Milkman KL et al (2017) Should governments invest more in nudging? Psychol Sci 28:1031–1040. https://doi.org/10.1177/2F0956797617702501

Bentham J (1789) An introduction to the principles of morals and legislation. Dover Publications, Mineola NY

Bowie NE (1999) Business ethics: a Kantian perspective. Blackwell Publishers, Malden, MA

Brey P (2011) Anticipatory technology ethics for emerging IT. In: Mauger J (ed) CEPE 2011: crossing boundaries. INSEIT, Nice, France, pp 13–26

Bynum T (2008a) Computer and information ethics. In: Zalta EN (ed) Stanford Encyclopedia of Philosophy. https://plato.stanford.edu/archives/fall2008/entries/ethics-computer/

Bynum TW (2006) Flourishing ethics. Ethics Inf Technol 8:157–173

Bynum TW (2008b) Norbert Wiener and the rise of information ethics. In: van den Hoven J, Weckert J (eds) Information technology and moral philosophy, 1st edn. Cambridge University Press, Cambridge, pp 8–25

Bynum TW, Rogerson S (2003) Computer ethics and professional responsibility: introductory text and readings. Blackwell Publishers, Cambridge, UK

Calás MB, Smircich L (1999) Past postmodernism? reflections and tentative directions. Acad Manage Rev 24:649–671. https://doi.org/10.2307/259347

Camerer CF, Loewenstein G, Rabin M (2004) Advances in behavioral economics. Princeton University Press, Princeton, NJ

Capurro R (2006) Towards an ontological foundation of information ethics. Ethics Inf Technol 8:175–186. https://doi.org/10.1007/s10676-006-9108-0

Cath CJN, Wachter S, Mittelstadt B, Taddeo M, Floridi L (2016) Artificial intelligence and the "good society": the US, EU, and UK approach. Social Science Research Network, Rochester, NY

Cecez-Kecmanovic D (2011) Doing critical information systems research: arguments for a critical research methodology. Eur J Inf Syst 20:440–455. https://doi.org/10.1057/ejis.2010.67

Clouser KD, Gert B (1990) A critique of principlism. J Med Philos 15:219–236. https://doi.org/10.1093/jmp/15.2.219

Coeckelbergh M (2019) Artificial intelligence: some ethical issues and regulatory challenges. In: Technology and Regulation, pp 31–34. https://doi.org/10.26116/techreg.2019.003

Council of Europe (1997) The Oviedo Convention: protecting human rights in the biomedical field. https://www.coe.int/en/web/bioethics/oviedo-convention. Accessed 30 Oct 2018

Craglia M, Annoni A, Benczur P et al (2018) Artificial intelligence: a European perspective. Publications Office of the European Union, Luxembourg

Creemers R (2018) China's social credit system: an evolving practice of control. Social Science Research Network, Rochester, NY

Dignum V (2019) Responsible artificial intelligence: how to develop and use AI in a responsible way. Springer Nature Switzerland AG, Cham, Switzerland

European Commission (2013) Options for strengthening responsible research and innovation. Publications Office of the European Union, Luxembourg

European Parliament (2020) Draft report with recommendations to the Commission on a framework of ethical aspects of artificial intelligence, robotics and related technologies. European Parliament, Committee on Legal Affairs. https://www.europarl.europa.eu/doceo/document/JURI-PR-650508_EN.pdf. Accessed 25 Sept 2020

European Union (2014) Rome declaration on responsible research and innovation in Europe https://ec.europa.eu/research/swafs/pdf/rome_declaration_RRI_final_21_November.pdf. Accessed 24 Sept 2020

Executive Office of the President (2016) Artificial intelligence, automation, and the economy. Executive Office of the President of the United States. https://obamawhitehouse.archives.gov/sites/whitehouse.gov/files/documents/Artificial-Intelligence-Automation-Economy.PDF. Accessed 23 Sept 2020

Feenberg A (1993) Critical theory of technology, New edn. Oxford University Press Inc, New York

Feenberg A (1999) Questioning technology, 1st edn. Routledge, London

Floridi L (2006) Information ethics, its nature and scope. ACM SIGCAS Comput Soc 36:21–36

Freyhofer HH (2004) The Nuremberg medical trial: the Holocaust and the origin of the Nuremberg Medical Code, 2nd revised edn. Peter Lang Publishing Inc, New York

Fuchs C, Mosco V (2017) Marx and the political economy of the media : studies in critical social science, reprint edn, vol 79. Haymarket Books, Chicago

Gilligan C (1990) In a different voice: psychological theory and women's development, reissue. Harvard University Press, Cambridge, MA

Greenhill A, Wilson M (2006) Haven or hell? Telework, flexibility and family in the e-society: a Marxist analysis. Eur J Inf Syst 15:379–388

Heidegger M (1993) Sein und Zeit, 14th edn. Max Niemeyer Verlag GmbH & Co KG, Tübingen

High-Level Expert Group on Artificial Intelligence (2019) Ethics guidelines for trustworthy AI. European Commission, Brussels. https://ec.europa.eu/newsroom/dae/document.cfm?doc_id=60419. Accessed 25 Sept 2020

House of Lords (2018) AI in the UK: ready, willing and able? HL Paper 100. Select Committee on Artificial Intelligence, House of Lords, Parliament, London. https://publications.parliament.uk/pa/ld201719/ldselect/ldai/100/100.pdf. Accessed 23 Sept 2020

Isaak J, Hanna MJ (2018) User data privacy: Facebook, Cambridge Analytica, and privacy protection. Computer 51:56–59. https://doi.ieeecomputersociety.org/10.1109/MC.2018.3191268

Jobin A, Ienca M, Vayena E (2019) The global landscape of AI ethics guidelines. Nat Mach Intell 1:389–399. https://doi.org/10.1038/s42256-019-0088-2

Johnstone J (2007) Technology as empowerment: a capability approach to computer ethics. Ethics Inf Technol 9:73–87

Kant I (1788) Kritik der praktischen Vernunft. Reclam, Ditzingen, Germany

Kant I (1797) Grundlegung zur Metaphysik der Sitten. Reclam, Ditzingen, Germany

Klar R, Lanzerath D (2020) The ethics of COVID-19 tracking apps: challenges and voluntariness. In: Research ethics. https://doi.org/10.1177/2F1747016120943622

Kleine D (2010) ICT4WHAT? Using the choice framework to operationalise the capability approach to development. J Int Dev 22:674–692. https://doi.org/10.1002/jid.1719

Klitzman R (2015) The ethics police? The struggle to make human research safe, 1st edn. Oxford University Press, New York

Liu C (2019) Multiple social credit systems in China. Social Science Research Network, Rochester, NY

Lyon D (2001) Surveillance society: monitoring everyday life. Open University Press, Buckingham, UK

MacIntyre AC (2007) After virtue: a study in moral theory. University of Notre Dame Press, Notre Dame, IN

Mayasandra R, Pan SL, Myers MD (2006) Viewing information technology outsourcing organizations through a postcolonial lens. In: Trauth E, Howcroft D, Butler T et al (eds) Social inclusion: societal and organizational implications for information systems. Springer Science+Business Media, New York, pp 381–396

Mill JS (1861) Utilitarianism, 2nd revised edn. Hackett Publishing Co, Indianapolis

Mullainathan S, Thaler RH (2000) Behavioral economics. NBER Working Paper No. 7948. National Bureau of Economic Research, Cambridge MA

Myers MD, Klein HK (2011) A set of principles for conducting critical research in information systems. MIS Q 35:17–36

National Commission for the Protection of Human Subjects of Biomedical and Behavioral Research (1979) The Belmont Report: ethical principles and guidelines for the protection of human subjects of research. US Government Printing Office, Washington DC

Nussbaum MC (2011) Creating capabilities: the human development approach. Harvard University Press, Cambridge, MA

Oosterlaken I, van den Hoven J (eds) (2012) The capability approach, technology and design. Springer, Dordrecht, Netherlands

Owen R (2014) The UK Engineering and Physical Sciences Research Council's commitment to a framework for responsible innovation. J Responsib Innov 1:113–117. https://doi.org/10.1080/23299460.2014.882065

Owen R, Pansera M (2019) Responsible innovation and responsible research and innovation. In: Simon D, Kuhlmann S, Stamm J, Canzler W (eds) Handbook on science and public policy. Edgar Elgar, Cheltenham UK, pp 26–48

Ryan M, Stahl BC (2020) Artificial intelligence ethics guidelines for developers and users: clarifying their content and normative implications. J Inf Commun Ethics Soc. https://doi.org/10.1108/JICES-12-2019-0138

Schrag ZM (2010) Ethical imperialism: institutional review boards and the social sciences, 1965–2009, 1st edn. Johns Hopkins University Press, Baltimore, MD

Schroeder D, Chatfield K, Singh M et al (2019) Equitable research partnerships: a global code of conduct to counter ethics dumping. Springer Nature, Cham, Switzerland

Sen A (2009) The idea of justice. Allen Lane, London

Stahl BC (2008) The ethical nature of critical research in information systems. Inf Syst J 18:137–163. https://doi.org/10.1111/j.1365-2575.2007.00283.x

Stahl BC (2012) Morality, ethics, and reflection: a categorization of normative IS research. J Assoc Inf Syst 13:636–656. https://doi.org/10.17705/1jais.00304

Stark L (2011) Behind closed doors: IRBs and the making of ethical research, 1st edn. University of Chicago Press, Chicago

Stilgoe J, Owen R, Macnaghten P (2013) Developing a framework for responsible innovation. Res Policy 42:1568–1580. https://doi.org/10.1016/j.respol.2013.05.008

Vallor S (2016) Technology and the virtues: a philosophical guide to a future worth wanting. Oxford University Press, New York

Van de Poel I, Royakkers L (2011) Ethics, technology, and engineering: an introduction. Wiley-Blackwell, Chichester, UK

van den Hoven J (2010) The use of normative theories in computer ethics. In: Floridi L (ed) The Cambridge handbook of information and computer ethics. Cambridge University Press, UK, pp 59–76

Von Schomberg R (2013) A vision of responsible research and innovation. In: Owen R, Heintz M, Bessant J (eds) Responsible innovation. Wiley, Chichester, UK, pp 51–74

Wiener N (1954) The human use of human beings. Doubleday, New York

Wiggershaus R (1995) The Frankfurt School: its history, theory and political significance, New edn. Polity Press, London

World Medical Association (2008) Declaration of Helsinki: ethical principles for medical research involving human subjects. https://web.archive.org/web/20091016152009/http://www.wma.net/en/30publications/10policies/b3/ Accessed 24 Sept 2020

Zuboff PS (2019) The age of surveillance capitalism: the fight for a human future at the new frontier of power. Profile Books, London

Chapter 4
Ethical Issues of AI

Abstract This chapter discusses the ethical issues that are raised by the development, deployment and use of AI. It starts with a review of the (ethical) benefits of AI and then presents the findings of the SHERPA project, which used case studies and a Delphi study to identify what people perceived to be ethical issues. These are discussed using the categorisation of AI technologies introduced earlier. Detailed accounts are given of ethical issues arising from machine learning, from artificial general intelligence and from broader socio-technical systems that incorporate AI.

Keywords Ethical issues of AI · Empirical ethics · Ethics and machine learning · Ethics of digital society · Metaphysical issues

Human flourishing as the foundation of AI ethics has provided the foundational basis for this book. We are now well equipped to explore ethical concerns in practice. This means that we now move from the conceptual to the empirical. In a first step I will give an overview of ethical issues, which I will then categorise in line with the earlier categorisation of concepts of AI.

4.1 Ethical Benefits of AI

When we speak of ethical issues of AI, there tends to be an implicit assumption that we are speaking of morally bad things. And, of course, most of the AI debate revolves around such morally problematic outcomes that need to be addressed. However, it is worth highlighting that AI promises numerous benefits. As noted earlier, many AI policy documents focus on the economic benefits of AI that are expected to arise from higher levels of efficiency and productivity. These are ethical values insofar as they promise higher levels of wealth and wellbeing that will allow people to live better lives and can thus be conducive to or even necessary for human flourishing. It is worth pointing out that this implies certain levels of distribution of wealth and certain assumptions about the role of society and the state in redistributing wealth in

© The Author(s) 2021

B. C. Stahl, *Artificial Intelligence for a Better Future*,
SpringerBriefs in Research and Innovation Governance,
https://doi.org/10.1007/978-3-030-69978-9_4

ethically acceptable manners which should be made explicit. The EU's High-Level Expert Group on AI (2019: 4) makes this very clear when it states:

> AI is not an end in itself, but rather a promising means to increase human flourishing, thereby enhancing individual and societal well-being and the common good, as well as bringing progress and innovation.

AI offers several other technical capabilities that can have immediate ethical benefits. The International Risk Governance Center (2018) names AI's analytical prowess, i.e. the ability to analyse quantities and sources of data that humans simply cannot process. AI can link data, find patterns and yield outcomes across domains and geographic boundaries. AI can be more consistent than humans, quickly adapt to changing inputs and free humans from tedious or repetitive tasks. These are all examples of technical capabilities that can easily be understood as being conducive to human flourishing because they lead to a better understanding and deeper insights into various phenomena. For instance, reducing commuting times or increasing the effectiveness of email spam filters are two everyday examples of where AI can make the life of busy professionals easier (Faggella 2020).

In addition to these examples of incidental ethical benefits, i.e. benefits that arise as a side effect of the technical capabilities of AI, there are increasing attempts to utilise AI specifically for ethical purposes. This is currently done under the heading of "AI for Good" (Berendt 2019). The key challenge that AI for Good faces is to define what counts as (ethically) good. In a pluralistic world there may often not be much agreement on what is good or why it would be considered good. However, there have been numerous attempts (e.g. Holmes et al. 2011) to identify shared ethical goods or values, such as benevolence, security, achievement and self-direction.

One can observe two different approaches to identifying the ethical goods that AI would have to promote to count as AI for Good: substantive goods and procedures to achieve them. Substantive goods are those practical outcomes that are universally, or at least very broadly, accepted to be good. The dominant example of such substantive moral goods is the UN's Sustainable Development Goals (SDGs) (Griggs et al. 2013). This set of 17 overarching goals has been described as "the world's best plan to build a better world for people and our planet" (United Nations 2020). It arose from decades of discussion of development policy and sustainability and evolved from the UN's Millenium Development Goals (Sachs 2012). The SDGs are interesting from an AI ethics perspective because they can be understood as the closest thing to humanity's consensus in terms of moral aims. They have been adopted by the UN and most member states and now have a pervasive presence in ethical debates. In addition, they are not only aspirational, but broken down into targets and measured by indicators and reported on by the UN and member states annually. It is therefore not surprising that one of the most visible attempts to promote AI for Good by the UN's International Telecommunications Union, the AI for Good Global Summit series,[1] has the strapline "Accelerating the United Nations Sustainable Development Goals".

While the SDGs are one dominant measure of the ethical benefit of AI, it is worth highlighting that they are not the only moral goods on which there is broad agreement.

---

[1] https://aiforgood.itu.int/.

Another huge body of work that indicates broad global agreement is built around human rights (Latonero 2018). Just like the SDGs, these were developed by the UN and codified. In addition, human rights have in many cases become enforceable through national legislation and in local courts. Upholding human rights is a condition of human flourishing (Kleinig and Evans 2013)

SDGs and human rights are two ways of determining the ethical benefits of AI. They therefore play a crucial role in the discussion of how ethical benefits and issues can be balanced, as I will show in more detail below when we come to the discussion of how ethical issues can be addressed.

4.2   Empirical Accounts of Ethical Issues of AI

There are numerous accounts of the ethical issues of AI, mostly developments of a long-standing tradition of discussing ethics and AI in the literature (Coeckelbergh 2019, Dignum 2019, Müller 2020), but increasingly also arising from a policy perspective (High-Level Expert Group on AI 2019). In this book and the SHERPA project[2] that underpins much of the argument, the aim was to go beyond literature reviews and find out empirically what people have in mind when they speak of the ethical issues of AI. I will focus here on ten case studies and the open-ended first stage of a Delphi study to come to a better understanding of how the ethics of AI is perceived by people working with and on AI systems.

The level of analysis of the case studies was defined as organisations that make use of AI. Case studies are a methodology that is recommended to provide answers to the "how" and "why" of a phenomenon and events over which the researcher has little or no control (Yin 2003a, b). In order to gain a broad understanding, a set of application areas of AI was defined and case study organisations identified accordingly. Using this methodology, the case studies covered the following social domains:

- employee monitoring and administration
- government
- agriculture
- sustainable development
- science
- insurance
- energy and utilities
- communications, media and entertainment
- retail and wholesale trade
- manufacturing and natural resources

For each case a minimum of two organisational members were interviewed, the aim being to engage with at least one technical expert who understood the system and one respondent with managerial or organisational expertise. Overall, for the ten

---

[2]https://www.project-sherpa.eu/

case studies, 42 individuals were interviewed. Based on the initial draft report of each case, peer review among the research team was undertaken, to ensure that the cases were consistent and comparable. For a detailed overview of the methods, findings and outcomes of the case study research, see Macnish et al. (2019)

The second piece of research that informs this chapter was the first stage of a three-stage Delphi study. Delphi studies are a well-established methodology to find solutions to complex and multi-faceted problems (Dalkey et al. 1969, Adler and Ziglio 1996, Linstone and Turoff 2002). They are typically expert-based and are used to find consensus among an expert population concerning a complex issue and to produce advice to decision-makers. Delphi studies normally involve several rounds of interaction, starting with broad and open questions, which are then narrowed down and prioritised.

The overview of ethical issues of AI that informs my discussion draws from the responses to the question in the first round of our Delphi Study. This was sent out to 250 experts on ethics and AI, selected from a range of stakeholders including technical experts, industry representatives, policymakers and civil society groups. Of these, 93 engaged with the online survey. A total of 41 usable responses were analysed. The open-ended question that was asked was: "What do you think are the three most important ethical or human rights issues raised by AI and/or big data?"

The analysis and findings of the first round were published and shared with the Delphi participants (Santiago 2020). These findings were then combined with the ones arrived at from the case study data analysis. Through group discussions similar relevant issues were combined and given suitable names or labels to ensure they were distinct and recognisable. For each of them a brief one-paragraph definition was provided.

The following list enumerates all the ethical issues that were identified from the case studies and the Delphi study, totalling 39.

1. Cost to innovation
2. Harm to physical integrity
3. Lack of access to public services
4. Lack of trust
5. "Awakening" of AI
6. Security problems
7. Lack of quality data
8. Disappearance of jobs
9. Power asymmetries
10. Negative impact on health
11. Problems of integrity
12. Lack of accuracy of data
13. Lack of privacy
14. Lack of transparency
15. Potential for military use
16. Lack of informed consent
17. Bias and discrimination

18. Unfairness
19. Unequal power relations
20. Misuse of personal data
21. Negative impact on justice system
22. Negative impact on democracy
23. Potential for criminal and malicious use
24. Loss of freedom and individual autonomy
25. Contested ownership of data
26. Reduction of human contact
27. Problems of control and use of data and systems
28. Lack of accuracy of predictive recommendations
29. Lack of accuracy of non-individual recommendations
30. Concentration of economic power
31. Violation of fundamental human rights in supply chain
32. Violation of fundamental human rights of end users
33. Unintended, unforeseeable adverse impacts
34. Prioritisation of the "wrong" problems
35. Negative impact on vulnerable groups
36. Lack of accountability and liability
37. Negative impact on environment
38. Loss of human decision-making
39. Lack of access to and freedom of information

There are several observations that could be made about this list. While in most cases one might intuitively accept that the issues can be seen as ethically relevant, no context or reason is provided as to why they are perceived to be ethically problematic. Many of them are not only ethically problematic but also directly linked to regulation and legislation. Being an ethical issue thus clearly does not exclude a given concern from being a legal issue at the same time.

The ethical issues are furthermore highly diverse in their specificity and likelihood of occurrence. Some are certain to come to pass, such as issues around data protection or data accuracy. Others are conceivable and likely, such as misuse or lack of trust. Yet others are somewhat diffuse, such as a negative impact on democracy, or on justice. In some cases, it is easy to see who should deal with the issues, while in others this is not so clear. This one-dimensional list of ethical issues is thus interesting as a first overview, but it needs to be processed further to be useful in considering how these issues can be addressed and what the priorities are.

It is possible to map the ethical issues to the different meanings of the concept of AI as outlined in Figure 2.1, as many of the issues are linked to the features of the different meanings as highlighted in Figure 2.2. I therefore distinguish three different sets of ethical issues: those arising from machine learning, general issues related to living in a digital world, and metaphysical issues (see Fig. 4.1).

Figure 4.1 indicates the relationship between the different categories of AI introduced in Chapter 2 and the ethical issues that will be discussed in the upcoming section. This relationship is indicative and should be understood as heuristic,

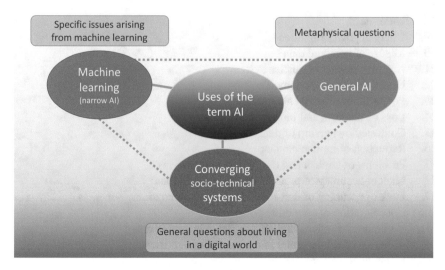

Fig. 4.1   Concepts of AI and ethical questions they raise

i.e. problem-solving, as the relationships in practice are more complex and not necessarily as linear as the figure might suggest.

4.3   Ethical Issues Arising from Machine Learning

The first set of issues consists of those that arise from the features of machine learning. Many of the machine learning techniques that led to the current success of AI are based on artificial neural networks. The features of these approaches that give rise to ethical concerns are opacity, unpredictability and the need for large datasets to train the technologies. Neither the developer, the deployer nor the user (see box) can normally know in advance how the system will react to a given set of inputs. And because the system learns and is thus adaptive and dynamic, past behaviours are not a perfect predictor for future behaviour in identical situations.

Developer, Deployer and User

Most current AI policy work distinguishes between developers, deployers and users (European Parliament 2020). The developer is the technical expert (or organisation) who builds the system. The deployer is the one who decides its use and thus has control over risks and benefits. In the case of an autonomous vehicle, for example, the developer might be the car manufacturer, and the deployer might be an organisation offering mobility services. A user is the one benefiting from the services. These roles may coincide, and a developer may be a deployer. Making the distinction seems reasonable, however, because a developer can be expected to have detailed understanding of the underlying technology, whereas the deployer may have much less insight.

A primary and frequently cited ethical issue is that of privacy and data protection. Privacy and data protection are not identical (Buttarelli 2018), but for the purposes of AI ethics, the key privacy concern is informational privacy, and data protection can be understood as a means to safeguard informational privacy. AI based on machine learning poses several risks to data protection. On the one hand it needs large data sets for training purposes, and the access to those data sets can raise questions of data protection. More interesting, and more specific to AI, is the problem that AI and its ability to detect patterns may pose privacy risks, even where no direct access to personal data is possible. The classic study by Jernigan and Mistree (2009) claiming to be able to identify sexual orientation from Facebook friendships is a good example. Notwithstanding the ethical and scientific merits of this particular study, it is easy to see that AI can be used to generate insights that raise privacy concerns. AI also has the potential of allowing the re-identification of anonymised personal data in ways that were not foreseen before the capabilities of machine learning became apparent. While data protection law is well established in most jurisdictions, AI has the potential to create new data protection risks not envisaged by legislation and thereby create new ethical concerns. AI may also use or generate types of personal data currently less widely employed, such as emotional personal data, further exacerbating the situation (Tao et al. 2005, Flick 2016).

Data protection concerns are directly linked to questions of data security. Cyber-security is a perennial problem of ICT, not just AI. However, AI systems may be subject to new types of security vulnerabilities, such as model poisoning attacks (Jagielski et al. 2018). Furthermore, these systems may be used for new types of vulnerability detection and exploitation (Krafft et al. 2020).

Privacy and data protection issues thus point to broader questions of reliability in AI systems. While reliability is a concern for all technical artefacts, the opacity of machine learning systems and their unpredictability mean that traditional deterministic testing regimes may not be applicable to them. The outputs of machine learning systems depend on the quality of the training data, which may be difficult to ascertain. The integrity of data can be threatened by security breaches, but also by technical or organisational aspects. This means that the reliability of machine learning systems may need to be assessed in different ways from other types of systems, which can be an ethical issue, if the system's output affects ethical value. For example, an AI system used for the identification of disease markers in pathology may work well under research conditions, with well-labelled training data, and perform at the level of a trained pathologist, or even better, under such conditions. This does not guarantee that the same system using the same model would perform as well under clinical conditions, which may be one of the reasons why, despite the great promise that AI holds for medicine, there are relatively few AI systems already in clinical practice (Topol 2019).

Machine learning systems are by definition not transparent, or at least not transparent in the way that other ICT systems could be. Where they are proprietary systems, the commercial confidentiality of algorithms and models may further limit transparency. "Transparency" is itself a contested term, but lack of transparency raises questions of accountability (USACM 2017). Lack of transparency makes it more difficult to recognise and address questions of bias and discrimination.

Bias is a much-cited ethical concern related to AI (CDEI 2019). One key challenge is that machine learning systems can, intentionally or inadvertently, result in the reproduction of already existing biases. There are numerous high-profile accounts of such cases, for example when gender biases in recruitment are replicated through the use of machine learning or when racial biases are perpetuated through machine learning in probation processes (Raso et al. 2018). Discrimination on the basis of certain (sometimes so-called protected) characteristics is not just an ethical issue but has long been recognised as a human rights infringement, and such discrimination therefore tends to be illegal in many jurisdictions. As AI poses a risk to this human right, there has been a focus on highlighting the potential of machine learning to infringe the right to equality and non-discrimination (Access Now Policy Team 2018).

Safety is also a key ethical issue of machine learning, in particular in systems that interact directly with the physical world, such as autonomous vehicles (BmVI 2017) or systems governing critical healthcare provision. While currently not very visible in the public debate, safety is sure to emerge prominently when machine-learning-enabled systems start to physically engage with humans more broadly.

The ethical issues set out in this section are directly related to the technical characteristics of machine learning. There are, however, numerous other ethical concerns which are less clearly linked to machine learning, many of which have to do with the characteristics of broader socio-technical systems that are discussed in the next section.

4.4   General Issues Related to Living in a Digital World

The second set of ethical issues consists of those that relate to what I called "AI as converging socio-technical systems". In Section 2.3 I suggested that these systems have the characteristics of autonomy, social impact and manipulation. To be clear, the distinction is an analytical one, as the converging socio-technical systems are not separate from machine learning systems but tend to include these and be based on machine learning *and* other AI capabilities. The difference is more one of perspective, where the term "machine learning" is used to focus on specific technologies for defined applications, whereas the converging socio-technical systems tend to involve numerous technologies and their focus is on the societal impact they cause.

I have chosen the label "living in a digital world" to describe these issues, in order to make it clear that most of them, while linked to AI, are not necessarily confined to AI. These questions are linked to broader societal and political decisions on how to structure and use large socio-technical systems. They can therefore not be viewed in separation from their societal role, and many of the ethical issues are directly caused by the way in which society and its actors work with these technologies.

An initial set of issues that arise from living in a digital world is related to the economy. The most prominent among these is likely to concern (un)employment. The potential of AI-related technologies to create a new wave of automation and thereby replace jobs has long been recognised (Collins 1990). In fact, Norbert Wiener suggested that computers competing with humans for jobs would have dire consequences for employment: "It is perfectly clear that this will produce an unemployment situation, in comparison with which the present recession and even the depression of the thirties will seem a pleasant joke" (Wiener 1954: 162).

While this bleak prediction has not (yet) come to pass, it is feared that AI will negatively affect employment. The novelty in the perceived threat from AI, which differs from earlier similar fears about ICT in general or other automation technologies, is that the jobs currently under apparent threat are better-paying ones: AI may increasingly imperil the income of middle-class professionals (Boden 2018). Losing employment is of course not only an economic problem; it also has social and psychological aspects (Kaplan and Haenlein 2019). The actual consequences of the introduction of AI for the employment market are at least partly an empirical question. The outcomes may be other than expected: jobs may not disappear but change instead (AI Now Institute 2017), and new jobs may be created, which may lead to new questions of fairness and distribution (House of Lords 2018).

The economic impacts of AI are not limited to employment. A further key concern is the concentration of economic (and by implication political) power. The reliance of current AI systems on large computing resources and massive amounts of data means that those organisations that own or have access to such resources are well placed to benefit from AI. The international concentration of such economic power among the big tech companies is independent of AI, but AI-related technologies have the potential to exacerbate the problem (Nemitz 2018).

These changes may not only be quantitative, i.e. related to the ability of large companies to make even more profits than they did prior to the use of AI, but may also be qualitatively different. Zuboff's (2019) concept of "surveillance capitalism" aims to capture the fundamental shifts in the economy that are facilitated by AI and the use of big data for behavioural prediction. Her argument is that these developments raise questions of fairness when large companies exploit user data that has been expropriated from individuals without compensation. The economic performance of large internet companies that make heavy use of AI certainly gives pause for thought. At the time of writing, Apple had just been valued as the most valuable global company, reaching a market value of $2 trillion. The stock market value of the five big internet companies – Apple, Microsoft, Amazon, Alphabet and Facebook – increased by $3 trillion during the COVID-19 pandemic, between 23 March and 19 August 2020 (Nicas 2020). This development may have more to do with the pathologies of the stock market than anything else, but it clearly shows that investors have huge hopes for the future of these companies – hopes that are likely to be related to their ability to harness AI.

Notwithstanding these astonishing figures, probably an even more important problem is that such companies utilise their insights to structure the space of action of individuals, thereby reducing the average citizen's ability to make autonomous choices. Such economic issues are thus directly related to broader questions of justice and fairness. There are immediate questions, such as the ownership of data and how this translates into the possibility of making use of the benefits of new technologies. Intellectual property has been a hotly debated topic in the ethics of computing for a long time (Spinello and Tavani 2004) and is now spilling over into the AI debate.

Another hotly debated issue is that of access to justice in the legal sense and how AI will transform the justice system. The use of AI for predictive policing or criminal probation services can broaden existing biases and further disadvantage parts of the population (Richardson et al. 2019).

While the use of AI in the criminal justice system may be the most hotly debated issue, AI is also likely to have impacts on access to other services, thereby potentially further excluding segments of the population that are already excluded. AI can thus exacerbate another well-established ethical concern of ICT, namely the so-called digital divide(s) (McSorley 2003, Parayil 2005, Busch 2011). Well-established categories of digital divides, such as the divides between countries, genders and ages, and between rural and urban, can all be exacerbated due to AI and the benefits it can create. These benefits imply that a lack of ability to access the underlying technology leads to missed opportunities, which can be an ethical concern.

Another basic category of ethical issues in the digital world is that of freedom. It is easy to see how the freedom of an individual whose parole decision was made or influenced by AI would be affected. However, the influence of AI on freedom is broader and more subtle. By providing or withdrawing access to information the technologies that surround us shape the space of possible action. The argument goes beyond Lessig's (1999) point that ICT is a form of law that allows or disallows certain actions. ICT in general and AI in particular can make a human's options appear or disappear without that human being aware of it. This does not even have to imply a conscious desire to mislead or deceive, but is simply an expression of the fact that our social reality is technically mediated and this mediation has consequences. An example would be the results of an internet search engine. Search engines rely heavily on AI. They also structure what users can see and will thus perceive as relevant, and how they then act. Search engine providers use this as part of their business model, by displaying paid-for content more prominently and enticing users to purchase. The point is, however, that even without such conscious attempts to direct users' attention, a search engine would still structure users' perception of reality and thus their scope of action.

As in the other cases, this is not necessarily negative. AI can open up enormous opportunities and create spaces for actions that were previously unthinkable, for example by allowing partially sighted people to drive vehicles autonomously, or by creating personalised medical solutions beyond what is currently possible. But at the same time, it can reduce individual autonomy, removing the freedom to decide and act in more or less subtle ways. An example might be the use of AI to steer visitors to a city on routes that avoid congestion and promote the safety of tourists (Ryan and

Gregory 2019). Such a system is based on morally desirable aims, but it still reduces the ability of individuals to move about the city as they would do in the absence of the system. This does not have to be an ethical issue, but it may have unintended consequences that are ethically problematic, for example when it reduces the footfall in parts of the city that depend on visitors.

Broader societal issues are power relationships and power asymmetries. Economic dominance and the structuring of options for action may give large amounts of power and dominance to some actors, to the point where democratic principles are jeopardised. The scandal around Facebook and Cambridge Analytica (Isaak and Hanna 2018) is a high-profile reminder of the potential vulnerabilities of democratic processes. But, as Coeckelbergh (2020: 100) points out, it is not just a problem of new forms of surveillance, manipulation and authoritarianism. Our democratic structures may be similarly undermined by "changing the economy in a way that turns us all into smartphone cattle milked for our data", thus linking back to Zuboff's pervasive theme of surveillance capitalism.

The list of possibly problematic issues of AI in different application areas is as long as the list of possible benefits. In most of these areas there are difficult questions about how to identify benefits and costs and what to do about them. A high-profile example is the use of AI for the creation of autonomous weapons. While it is easy to see that saving soldiers' lives by replacing them with robots would be an ethical benefit, there are numerous counterarguments ranging from the practical, such as the reliability of such systems, to the political, such as whether they would lower the threshold to starting wars, to the fundamental, such as whether it can ever by appropriate to take human lives on the basis of machine input (Sharkey 2017, Defense Innovation Board 2019, Babuta et al. 2020).

Similar worries arise in AI for health, where technology can improve diagnoses and treatments, but may have risks and downsides. An example would be care technologies: robotic systems have long been proposed as a way of addressing challenges faced by the care sector, but there are concerns about replacing human contact with technology, which is often seen as a fundamental ethical issue (Decker 2008, Sharkey and Sharkey 2010, Goeldner et al. 2015).

These broader societal issues are not confined to direct impact on human lives and actions, but also take in the impact of AI on the environment. While AI offers the possibility of decreased power consumption by streamlining processes, it simultaneously requires large amounts of resources and it creates new products and services that can have negative impacts on the environment.

4.5 Metaphysical Issues

This discussion of ethical issues of AI started with the most immediate issues arising from a specific technology, namely machine learning, and then progressed to broader societal concerns. The third and final category of ethical issues, what I call "metaphysical issues", is the most open and unexplored one. I have used the term "metaphysical" because the issues here are directly linked to fundamental aspects of reality, of the nature of being and human ability to make sense of this. They also go to the heart of the nature of humans and humanity.

These metaphysical issues are mostly related to artificial general intelligence (AGI) or good old-fashioned AI (GOFAI), which is typically conceptualised in terms of a symbolic and logical representation of the world. The idea is that AGI (which may build on GOFAI, but does not have to) would display human reasoning abilities. To reiterate a point made earlier: there currently are no AGI systems available, and there is considerable disagreement about their possibility and likelihood. I am personally not convinced that they are possible with current technologies, but I cannot prove the point any more definitively than others, so I will remain agnostic on the point of fundamental possibility. What seems abundantly clear, however, is that progress in the direction of AGI is exceedingly slow. Hence, I do not expect any technology that would be accepted as AGI by the majority of the expert communities to come into existence during the coming decades.

The metaphysical ethical issues raised by AGI are therefore not particularly urgent, and they do not drive policy considerations in the way that issues like discrimination or unemployment do. Most policy documents on AI ignore these issues, on the implicit assumption that they are not in need of policy development. In the empirical research presented earlier in this section, these metaphysical issues were not identified as issues that organisations currently engage with. There is probably also an element of fear on the part of scholars and experts of being stigmatised as not being serious or scholarly, as these metaphysical issues are the staple of science fiction.

I nevertheless include them in this discussion of ethical issues of AI for two reasons. Firstly, these questions are thought-provoking, not only for experts but for media and society at large, because they touch on many of the fundamental questions of ethics and humanity. Secondly, some of these issues can shed light on the practical issues of current AI by forcing clearer reflection on key concepts, such as autonomy and responsibility and the role of technology in a good society.

The techno-optimistic version of AGI is that there will be a point when AI is sufficiently advanced to start to self-improve, and an explosion of intelligence – the singularity (Kurzweil 2006) – will occur due to a positive feedback loop of AI onto itself. This will lead to the establishment of super-intelligence (Bostrom 2016). The implication is that AGI will then not only be better than humans at most or all cognitive tasks, but will also develop consciousness and self-awareness (Torrance 2012). The contributors to this discussion disagree on what would happen next. The super-intelligent AGI might be benevolent and make human life better, it might see humans as competitors and destroy us, or it might reside in a different sphere of consciousness, ignoring humanity for the most part.

Speculations along those lines are not particularly enlightening: they say more about the worldview of the speculator than anything else. But what is interesting is to look at some of the resulting ethical issues in light of current technologies. One key question is whether such AGIs could be subjects of responsibility. Could we hold them morally responsible for their actions or the consequences of these actions (Bechtel 1985)? To put it differently, is there such a thing as artificial morality (Wallach and Allen 2008, Wallach et al. 2011)? This question is interesting because it translates into the question: can we hold current AIs responsible? And this is a practical question in cases where AIs can create morally relevant consequences, as is the case for autonomous vehicles and many other systems that interact with the world.

The question whether an entity can be a subject of moral responsibility, i.e. someone or something of which or whom we can say, "X is responsible," hinges on the definition of responsibility (Fischer 1999). There is a large literature on this question, and responsibility subjects typically have to fulfil a number of requirements, which include an understanding of the situation, a causal role in events, the freedom to think and act, and the power to act, to give four examples.

The question of whether computers can be responsible is therefore somewhat similar to the question of whether they can think. One could argue that, if they can think, they can be responsible. However, Turing (1950) held the question of whether machines can think to be meaningless and proposed the imitation game, i.e. the Turing test, instead. In light of the difficulty of the question it is therefore not surprising that an analogous approach to machine responsibility was devised, the moral Turing test, where the moral status of a machine could be defined by the fact that it was recognised as a moral agent by an independent interlocutor. The problem with that approach is that it does not really address the issue. I have elsewhere suggested that a machine that can pass the Turing test could probably also pass a moral Turing test (Stahl 2004).

Much of the discussion of the moral status of AI hinges on the definition of "ethics". If one takes a utilitarian position, for example, it would seem plausible to assume that computers would be at least as good as humans at undertaking a moral calculus, provided they had the data to comprehensively describe possible states of the world. This seems to be the reason why the trolley problem is so prominent in the discussion of the ethics of autonomous vehicles (Wolkenstein 2018). The trolley problem,[3] which is based on the premise that an agent has to make a dilemmatic decision between two alternatives, either of which will typically kill different actors, has caught the attention of some scholars because it seems to map to possible real-world scenarios in AI, notably with regard to the programming or behaviour of autonomous vehicles. An autonomous vehicle can conceivably be put in a situation that is similar to the trolley problem in that it has to make a rapid decision between two ethically problematic outcomes. However, I would argue that this is based on a misunderstanding of the trolley problem, which was devised by Philippa Foot (1978) as an analytical tool to show the limitations of moral reasoning, in particular utilitarianism. The dilemma structure is geared towards showing that there is not one "ethically correct" response. It has therefore been argued (Etzioni and Etzioni 2017), rightly in my opinion, that the trolley problem does not help us determine whether

---

[3] A typical trolley problem would see an agent standing near the points where two railway lines merge into a single track. From the single track, a train approaches. Unaware of the train, a number of children are playing on the left-hand track, whereas a single labourer, also unaware of the train, is working on the right-hand track. The train is set to hit the children. By switching the points, the agent can switch the train onto the right-hand track, thereby saving the children's lives, but leading to a single death. What should the agent do? That is the trolley problem.

machines can be ethical, because it can fully be resolved with recourse to existing standards of human responsibility.

I have argued earlier that the key to understanding ethics is an understanding of the human condition. We develop and use ethics because we are corporeal, and hence vulnerable and mortal, beings who can feel empathy with others who have fears and hopes similar to our own. This is the basis of our social nature and hence of our ethics. If we use this starting point, then AI, in order to be morally responsible and an ethical agent, would have to share these characteristics. At the moment no system comes close to empathy. This has nothing to do with AI's computational abilities, which far exceed ours and have done for some time, but arises from the fact that AI is simply not in the same category as we are.

This does not mean that we cannot assign a moral status to AI, or to some type of AI. Humans can assign such a status to non-humans and have always done so, for example by viewing parts of nature or artefacts as divine or by protecting certain entities from being treated in certain ways.

Such a view of AI has the advantage of resolving some of the metaphysical questions immediately. If an existentialist commitment to our shared social world is a condition of being an ethical agent, then current AI simply falls out of the equation. This does not mean that developers of autonomous vehicles do not need to worry any more, but it does mean that they can use established mechanisms of responsibility, accountability and liability to make design decisions. It also does not fundamentally rule out artificial moral agents, but these would have to be of a very different nature from current computing technologies.

This position does not solve all metaphysical questions. There are interesting issues arising from the combination of humans and machines that need attention. Actor-networks containing AI-enabled artefacts may well change some of our ethical perceptions. The more AI gets integrated into our nature, the more it raises new questions. This starts with seemingly trivial aspects of the prevalence of ubiquitous devices such as mobile phones and what these do to our agency. Cutting-edge technologies, such as AI-supported brain computer interfaces, change what we can do, but they can also change how we ascribe responsibility. In this sense questions of posthumanism (Barad 2003) and human enhancement (Bostrom and Sandberg 2009, Coeckelbergh 2011) may be more interesting from the AI ethics perspective because they start with existing ethical agency that may need to be adjusted.

Much more could of course be said about ethical issues of AI, but this chapter has hopefully given a good overview and provided a useful categorisation of these issues, as shown in Table 4.1.

The categorisation in Table 4.1 is not authoritative, and others are possible. A different view that would come to similar conclusions would focus on the temporal nature of the issues. Ordering ethical issues of AI by temporal proximity and urgency is not new. Baum (2018) has suggested the distinction between "presentists" and "futurists", calling attention to near-term and long-term AI issues. Extending this thought to the discussion of ethical issues of AI as presented in this chapter, one can say that the ethical issues of machine learning are the most immediate ones and the metaphysical ones are long-term, if not perpetual, questions. The category of issues

Table 4.1 Three categories of ethical issues of artificial intelligence

1. Issues arising from machine learning	
Privacy and data protection	Lack of privacy
	Misuse of personal data
	Security problems
Reliability	Lack of quality data
	Lack of accuracy of data
	Problems of integrity
Transparency	Lack of accountability and liability
	Lack of transparency
	Bias and discrimination
	Lack of accuracy of predictive recommendations
	Lack of accuracy of non-individual recommendations
Safety	Harm to physical integrity
2. Living in a digital world	
Economic issues	Disappearance of jobs
	Concentration of economic power
	Cost to innovation
Justice and fairness	Contested ownership of data
	Negative impact on justice system
	Lack of access to public services
	Violation of fundamental human rights of end users
	Violation of fundamental human rights in supply chain
	Negative impact on vulnerable groups
	Unfairness
Freedom	Lack of access to and freedom of information
	Loss of human decision-making
	Loss of freedom and individual autonomy
Broader societal issues	Unequal power relations
	Power asymmetries
	Negative impact on democracy
	Problems of control and use of data and systems
	Lack of informed consent
	Lack of trust
	Potential for military use
	Negative impact on health
	Reduction of human contact
	Negative impact on environment
Uncertainty issues	Unintended, unforeseeable adverse impacts
	Prioritisation of the "wrong" problems
	Potential for criminal and malicious use
3. Metaphysical issues	
	Machine consciousness
	"Awakening" of AI
	Autonomous moral agents
	Super-intelligence
	Singularity
	Changes to human nature

arising from living in the digital world is located somewhere between. This view may also have implications for the question of how, when and by whom ethical issues in AI can be addressed, which will be discussed in the next chapter.

References

Access Now Policy Team (2018) The Toronto declaration: protecting the right to equality and non-discrimination in machine learning systems. Access Now, Toronto. https://www.access now.org/cms/assets/uploads/2018/08/The-Toronto-Declaration_ENG_08-2018.pdf. Accessed 26 Sept 2020

Adler M, Ziglio E (eds) (1996) Gazing into the oracle: the Delphi method and its application to social policy and public health. Jessica Kingsley, London

AI Now Institute (2017) AI Now 2017 report. https://ainowinstitute.org/AI_Now_2017_Report. pdf. Accessed 26 Sept 2020

Babuta A, Oswald M, Janjeva A (2020) Artificial intelligence and UK national security: policy considerations. RUSI Occasional Paper. Royal United Services Institute for Defence and Security Studies, London. https://rusi.org/sites/default/files/ai_national_security_final_web_version.pdf. Accessed 21 Sept 2020

Barad K (2003) Posthumanist performativity: toward an understanding of how matter comes to matter. Signs 28:801–831. https://doi.org/10.1086/345321

Baum SD (2018) Reconciliation between factions focused on near-term and long-term artificial intelligence. AI Soc 33:565–572. https://doi.org/10.1007/s00146-017-0734-3

Bechtel W (1985) Attributing responsibility to computer systems. Metaphilosophy 16:296–306. https://doi.org/10.1111/j.1467-9973.1985.tb00176.x

Berendt B (2019) AI for the common good?! Pitfalls, challenges, and ethics pen-testing. Paladyn J Behav Robot 10:44–65. https://doi.org/10.1515/pjbr-2019-0004

BmVI (2017) Ethik-Kommission: Automatisiertes und vernetztes Fahren. Bundesministerium für Verkehr und digitale Infrastruktur, Berlin. https://www.bmvi.de/SharedDocs/DE/Publikationen/DG/bericht-der-ethik-kommission.pdf?__blob=publicationFile. Accessed 26 Sept 2020

Boden MA (2018) Artificial intelligence: a very short introduction, Reprint edn. Oxford University Press, Oxford

Bostrom N (2016) Superintelligence: paths, dangers, strategies, Reprint edn. Oxford University Press, Oxford and New York

Bostrom N, Sandberg A (2009) Cognitive enhancement: methods, ethics, regulatory challenges. Sci Eng Ethics 15:311–341

Busch T (2011) Capabilities in, capabilities out: overcoming digital divides by promoting corporate citizenship and fair ICT. Ethics Inf Technol 13:339–353

Buttarelli G (2018) Choose humanity: putting dignity back into digital. In: Speech at 40th international conference of data protection and privacy commissioners, Brussels. https://www.privacyconference2018.org/system/files/2018-10/Choose%20Humanity%20speech_0.pdf. Accessed 26 Sept 2020

CDEI (2019) Interim report: Review into bias in algorithmic decision-making. Centre for Data Ethics and Innovation, London. https://www.gov.uk/government/publications/interim-reports-from-the-centre-for-data-ethics-and-innovation/interim-report-review-into-bias-in-algorithmic-decision-making. Accessed 26 Sept 2020

Coeckelbergh M (2011) Human development or human enhancement? A methodological reflection on capabilities and the evaluation of information technologies. Ethics Inf Technol 13:81–92. https://doi.org/10.1007/s10676-010-9231-9

Coeckelbergh M (2019) Artificial Intelligence: some ethical issues and regulatory challenges. In: Technology and regulation, pp 31–34. https://doi.org/10.26116/techreg.2019.003

Coeckelbergh M (2020) AI ethics. The MIT Press, Cambridge, MA

Collins HM (1990) Artificial experts: social knowledge and intelligent systems. MIT Press, Cambridge, MA

Dalkey NC, Brown BB, Cochran S (1969) The Delphi method: an experimental study of group opinion. Rand Corporation, Santa Monica, CA

Decker M (2008) Caregiving robots and ethical reflection: the perspective of interdisciplinary technology assessment. AI Soc 22:315–330

Defense Innovation Board (2019) AI principles: recommendations on the ethical use of artificial intelligence by the Department of Defense. US Department of Defense, Washington DC. https://media.defense.gov/2019/Oct/31/2002204458/-1/-1/0/DIB_AI_PRINCIPLES_PRIMARY_DOCUMENT.PDF. Accessed 26 Sept 2020

Dignum V (2019) Responsible artificial intelligence: how to develop and use AI in a responsible way. Springer Nature Switzerland AG, Cham, Switzerland

Etzioni A, Etzioni O (2017) Incorporating ethics into artificial intelligence. J Ethics 21:403–418. https://doi.org/10.1007/s10892-017-9252-2

European Parliament (2020) Draft report with recommendations to the Commission on a framework of ethical aspects of artificial intelligence, robotics and related technologies. European Parliament, Committee on Legal Affairs. https://www.europarl.europa.eu/doceo/document/JURI-PR-650508_EN.pdf. Accessed 25 Sept 2020

Faggella D (2020) Everyday examples of artificial intelligence and machine learning. Emerj, Boston, MA. https://emerj.com/ai-sector-overviews/everyday-examples-of-ai/. Accessed 23 Sep 2020

Fischer JM (1999) Recent work on moral responsibility. Ethics 110:93–139

Flick C (2016) Informed consent and the Facebook emotional manipulation study. Res Ethics 12. https://doi.org/10.1177/1747016115599568

Foot P (1978) Virtues and vices and other essays in moral philosophy. University of California Press, Berkeley and Los Angeles

Goeldner M, Herstatt C, Tietze F (2015) The emergence of care robotics: a patent and publication analysis. Technol Forecast Soc Change 92:115–131. https://doi.org/10.1016/j.techfore.2014.09.005

Griggs D, Stafford-Smith M, Gaffney O et al (2013) Sustainable development goals for people and planet. Nature 495:305–307. https://doi.org/10.1038/495305a

High-Level Expert Group on Artificial Intelligence (2019) Ethics guidelines for trustworthy AI. European Commission, Brussels. https://ec.europa.eu/newsroom/dae/document.cfm?doc_id=60419. Accessed 25 Sept 2020

Holmes T, Blackmore E, Hawkins R (2011) The common cause handbook: a guide to values and frames for campaigners, community organisers, civil servants, fundraisers, educators, social entrepreneurs, activists, funders, politicians, and everyone in between. Public Interest Research Centre, Machynlleth UK

House of Lords (2018) AI in the UK: ready, willing and able? HL Paper 100. Select Committee on Artificial Intelligence, House of Lords, Parliament, London. https://publications.parliament.uk/pa/ld201719/ldselect/ldai/100/100.pdf. Accessed 23 Sept 2020

International Risk Governance Center (IRGC) (2018) The governance of decision-making algorithms. EPFL International Risk Governance Center, Lausanne. https://infoscience.epfl.ch/record/261264/files/IRGC%20%282018%29%20The%20Governance%20of%20Decision-Making%20Algorithms-Workshop%20report.pdf. Accessed 26 Sept 2020

Isaak J, Hanna MJ (2018) User data privacy: Facebook, Cambridge Analytica, and privacy protection. Computer 51:56–59. https://doi.ieeecomputersociety.org/10.1109/MC.2018.3191268

Jagielski M, Oprea A, Biggio B et al. (2018) Manipulating machine learning: poisoning attacks and countermeasures for regression learning. In: 2018 IEEE symposium on security and privacy (SP), San Francisco. https://doi.org/10.1109/SP.2018.00057

Jernigan C, Mistree BFT (2009) Gaydar: Facebook friendships expose sexual orientation. First Monday 14. https://firstmonday.org/ojs/index.php/fm/article/download/2611/2302. Accessed 26 Sept 2020

Kaplan A, Haenlein M (2019) Siri, Siri, in my hand: who's the fairest in the land? On the interpretations, illustrations, and implications of artificial intelligence. Bus Horiz 62:15–25

Kleinig J, Evans NG (2013) Human flourishing, human dignity, and human rights. Law Philos 32:539–564. https://doi.org/10.1007/s10982-012-9153-2

Krafft T, Hauer M, Fetic L et al (2020) From principles to practice: an interdisciplinary framework to operationalise AI ethics. VDE and Bertelsmann Stiftung. https://www.ai-ethics-impact.org/res ource/blob/1961130/c6db9894ee73aefa489d6249f5ee2b9f/aieig—report—download-hb-data. pdf. Accessed 26 Sept 2020

Kurzweil R (2006) The singularity is near. Gerald Duckworth & Co, London

Latonero M (2018) Governing artificial intelligence: upholding human rights & dignity. Data & Society. https://datasociety.net/wp-content/uploads/2018/10/DataSociety_Governing_Artifi cial_Intelligence_Upholding_Human_Rights.pdf. Accessed 26 Sept 2020

Lessig L (1999) Code: and other laws of cyberspace. Basic Books, New York

Linstone HA, Turoff M (eds) (2002) The Delphi method: techniques and applications. Addison-Wesley Publishing Company, Advanced Book Program. https://web.njit.edu/~turoff/pubs/delphi book/delphibook.pdf. Accessed 26 Sept 2020

Macnish K, Ryan M, Gregory A et al. (2019) SHERPA deliverable D1.1 Case studies. De Montfort University. https://doi.org/10.21253/DMU.7679690.v3

McSorley K (2003) The secular salvation story of the digital divide. Ethics Inf Technol 5:75–87. https://doi.org/10.1023/A:1024946302065

Müller VC (2020) Ethics of artificial intelligence and robotics. In: Zalta EN (ed) The Stanford encyclopedia of philosophy, Fall 2020. Metaphysics Research Lab, Stanford University, Stanford, CA

Nemitz P (2018) Constitutional democracy and technology in the age of artificial intelligence. Phil Trans R Soc A 376:20180089. https://doi.org/10.1098/rsta.2018.0089

Nicas J (2020) Apple reaches $2 trillion, punctuating big tech's grip. The New York Times. https://www.nytimes.com/2020/08/19/technology/apple-2-trillion.html. Accessed 26 Sept 2020

Parayil G (2005) The digital divide and increasing returns: contradictions of informational capitalism. Inf Soc 21:41–51. https://doi.org/10.1080/01972240590895900

Raso FA, Hilligoss H, Krishnamurthy V et al. (2018) Artificial intelligence & human rights: opportunities & risks. Berkman Klein Center Research Publication No. 2018-6. http://dx.doi.org/10.2139/ssrn.3259344

Richardson R, Schultz J, Crawford K (2019) Dirty data, bad predictions: how civil rights violations impact police data, predictive policing systems, and justice. N Y Univ Law Rev Online 192. https://ssrn.com/abstract=3333423. Accessed 26 Sept 2020

Ryan M, Gregory A (2019) Ethics of using smart city AI and big data: the case of four large European cities. ORBIT J 2. https://doi.org/10.29297/orbit.v2i2.110

Sachs JD (2012) From millennium development goals to sustainable development goals. Lancet 379:2206–2211. https://doi.org/10.1016/S0140-6736(12)60685-0

Santiago N (2020) Shaping the ethical dimensions of smart information systems: a European perspective. SHERPA Delphi study, round 1 results. SHERPA project. https://www.project-she rpa.eu/wp-content/uploads/2020/03/sherpa-delphi-study-round-1-summary-17.03.2020.docx. pdf. Accessed 26 Sept 2020

Sharkey A, Sharkey N (2010) Granny and the robots: ethical issues in robot care for the elderly. Ethics Inf Technol. https://doi.org/10.1007/s10676-010-9234-6

Sharkey N (2017) Why robots should not be delegated with the decision to kill. Conn Sci 29:177–186. https://doi.org/10.1080/09540091.2017.1310183

Spinello RA, Tavani HT (2004) Intellectual property rights in a networked world: theory and practice. Information Science Publishing, Hershey PA

Stahl BC (2004) Information, ethics, and computers: the problem of autonomous moral agents. Minds Mach (Dordr) 14:67–83. https://doi.org/10.1023/B:MIND.0000005136.61217.93

Tao J, Tan T, Picard R (2005) Affective computing and intelligent interaction. Springer, Berlin

Topol EJ (2019) High-performance medicine: the convergence of human and artificial intelligence. Nat Med 25:44–56. https://doi.org/10.1038/s41591-018-0300-7

Torrance S (2012) Super-intelligence and (super-)consciousness. Int J Mach Conscious 4:483–501. https://doi.org/10.1142/S1793843012400288

Turing AM (1950) Computing machinery and intelligence. Mind 59:433–460. https://doi.org/10.1093/mind/LIX.236.433

United Nations (2020) Sustainable development knowledge platform. https://sustainabledevelopment.un.org. Accessed 25 May 2020

USACM (2017) Statement on algorithmic transparency and accountability. ACM US Public Policy Council, Washington DC. https://www.acm.org/binaries/content/assets/public-policy/2017_usacm_statement_algorithms.pdf. Accessed 26 Sept 2020

Wallach W, Allen C (2008) Moral machines: teaching robots right from wrong. Oxford University Press, New York

Wallach W, Allen C, Franklin S (2011) Consciousness and ethics: artificially conscious moral agents. Int J Mach Conscious 3:177–192. https://doi.org/10.1142/S1793843011000674

Wiener N (1954) The human use of human beings. Doubleday, New York

Wolkenstein A (2018) What has the trolley dilemma ever done for us (and what will it do in the future)? On some recent debates about the ethics of self-driving cars. Ethics Inf Technol 1–11. https://doi.org/10.1007/s10676-018-9456-6

Yin RK (2003a) Applications of case study research, 2nd edn. Sage Publications, Thousand Oaks CA

Yin RK (2003b) Case study research: design and methods, 3rd edn. Sage Publications, Thousand Oaks, CA

Zuboff PS (2019) The age of surveillance capitalism: the fight for a human future at the new frontier of power. Profile Books, London

Chapter 5
Addressing Ethical Issues in AI

Abstract This chapter reviews the proposals that have been put forward to address ethical issues of AI. It divides them into policy-level proposals, organisational responses and guidance for individuals. It discusses how these mitigation options are reflected in the case studies exemplifying the social reality of AI ethics. The chapter concludes with an overview of the stakeholder groups affected by AI, many of whom play a role in implementing the mitigation strategies and addressing ethical issues in AI.

Keywords AI ethics mitigation · AI policy · AI legislation · AI regulator · Organisational responses to AI · AI ethics guidance · AI ethics stakeholders

We now have a detailed, empirically informed and, I hope, conceptually interesting view of the ethics of AI. This leads to the question: what can we do about it? This chapter gives an overview of possible answers currently being discussed in the academic and policy discourses. For ease of reading, it breaks down the options into policy level, organisational level, guidance mechanisms and supporting activities. For each of these categories key mitigation measures will be introduced and key open issues and questions highlighted.

5.1 Options at the Policy Level

Activities at policy level are undertaken by political decision-makers. These can be located at the national level, but also at the regional and/or international level. Due to the nature of AI as an international and cross-boundary technology, particular attention will be paid to international policy initiatives coming from international bodies such as the UN, the United Nations Educational, Scientific and Cultural Organization (UNESCO) and the OECD. And, as a European writing a book based to a large extent on European research, I focus my attention mainly on European policy.

© The Author(s) 2021
B. C. Stahl, *Artificial Intelligence for a Better Future*,
SpringerBriefs in Research and Innovation Governance,
https://doi.org/10.1007/978-3-030-69978-9_5

5.1.1 Policy Aims and Initiatives

The number of policy papers on AI is significant. Jobin et al. (2019) have provided a very good overview, but it is no longer comprehensive, as the publication of policy papers continues unabated. Several individuals and groups have set up websites, databases, observatories or other types of resources to track this development. Some of the earlier ones seem to have been one-off overviews that are no longer maintained, such as the websites by Tim Dutton (2018), Charlotte Stix (n.d.) and NESTA (n.d.). Others remain up to date, such as the website run by AlgorithmWatch (n.d.), or have only recently come online, such as the websites by Ai4EU (n.d.), the EU's Joint Research Centre (European Commission n.d.) and the OECD (n.d.).

What most of these policy initiatives seem to have in common is that they aim to promote the development and use of AI, while paying attention to social, ethical and human rights concerns, often using the term "trustworthy AI" to indicate attention to these issues. A good example of high-level policy aims that are meant to guide further policy development is provided by the OECD (2019). It recommends to its member states that they develop policies for the following five aims:

- investing in AI research and development
- fostering a digital ecosystem for AI
- shaping an enabling policy environment for AI
- building human capacity
- preparing for labour market transformation
- international co-operation for trustworthy AI

Policy initiatives aimed at following these recommendations can cover a broad range of areas, most of which have relevance to ethical issues. They can address questions of access to data, distribution of costs and benefits through taxation or other means, environmental sustainability and green IT, to give some prominent examples.

These policy initiatives can be aspirational or more tangible. In order for them to make a practical difference, they need to be translated into legislation and regulation, as will be discussed in the following section.

5.1.2 Legislation and Regulation

At the time of writing this text (European summer 2020), there is much activity in Europe directed towards developing appropriate EU-level legislation and regulation of AI. The European Commission has launched several policy papers and proposals (e.g. European Commission 2020c, d), notably including a White Paper on AI (European Commission 2020a). The European Parliament has shared some counterproposals (European Parliament 2020a, b) and the political process is expected to lead to legislative action in 2021.

Using the categories developed in this book, the question is whether – for the purpose of the legislation – AI research, development and use will be framed in terms of human flourishing, efficiency or control. The EC's White Paper (European Commission 2020a) is an interesting example to use when studying the relationship between these different purposes. To understand this relationship, it is important to see that the EC uses the term "trust" to represent ethical and social aspects, following the High-Level Expert Group on AI (2019). This suggests that the role of ethics is to allow people to trust a technology that has been pre-ordained or whose arrival is inevitable. In fact, the initial sentences in the introduction to the White Paper state exactly that: "As digital technology becomes an ever more central part of every aspect of people's lives, people should be able to trust it. Trustworthiness is also a prerequisite for its uptake" (European Commission 2020a: 1). Ethical aspects of AI are typically discussed by European bodies using the terminology of trust. The document overall often follows this narrative and the focus is on the economic advantages of AI, including the improvement of the EU's competitive position in the perceived international AI race.

However, there are other parts of the document that focus more on the human flourishing aspect: AI systems are described as having a "significant role in achieving the Sustainable Development Goals" (European Commission 2020a: 2), environmental sustainability and ethical objectives. It is not surprising that a high-level policy initiative like the EC's White Paper combines different policy objectives. What is nevertheless interesting to note is that the White Paper contains two main areas of policy objectives: excellence and trust. In Section 4, entitled "An ecosystem of excellence", the paper lays out policies to strengthen the scientific and technical bases of European AI, covering European collaboration, research, skills, work with SMEs and the private sector, and infrastructure. Section 5, the second main part of the White Paper, under "An ecosystem of trust", focuses on risks, potential harms, liability and similar regulatory aspects. This structure of the White Paper can be read to suggest that excellence and trust are fundamentally separate, and that technical AI development is paramount, requiring ethics and regulation to follow.

When looking at the suitability of legislation and regulation to address ethical issues of AI, one can ask whether and to what degree these issues are already covered by existing legislation. In many cases the question thus is whether legislation is fit for purpose or whether it needs to be amended in light of technical developments. Examples of bodies of law with clear relevance to some of the ethical issues are intellectual property law, data protection law and competition law.

One area of law that is likely to be relevant and has already led to much high-level debate is that of liability law. Liability law is used to deal with risks and damage sustained from using (consumer) products, whether derived from new technologies or not. Liability law is likely to play a key role in distributing risks and benefits of AI (Garden et al. 2019). This explains the various EU-level initiatives (Expert Group on Liability and New Technologies 2019; European Commission 2020b; European Parliament 2020b) that try to establish who is liable for which aspects of AI. Relatedly, the allocation of strict and tort liabilities will set the scene for the greater AI environment, including insurance and litigation.

Another body of existing legislation and regulation being promoted to address the ethical issues of AI is that of human rights legislation. It has already been highlighted that many of the ethical issues of AI are simultaneously human rights issues, such as privacy and discrimination. Several contributors to the debate therefore suggest that existing human rights regulation may be well suited to addressing AI ethics issues. Proposals to this effect can focus on particular technologies, such as machine learning (Access Now Policy Team 2018), or on particular application areas, such as health (Committee on Bioethics 2019), or broadly propose the application of human rights principles to the entire field of AI (Latonero 2018, Commissioner for Human Rights 2019, WEF 2019).

The discussion of liability principles at EU level is a good example of the more specific regulatory options that are being explored. During a recent review of regulatory options for the legislative governance of AI, in particular at the European level, Rodrigues et al. (2020) surveyed the current legislative landscape and identified the following proposals that are under active discussion:

- the adoption of common EU definitions
- algorithmic impact assessments under the General Data Protection Regulation (GDPR)
- creating electronic personhood status for autonomous systems
- the establishment of a comprehensive EU system of registration of advanced robots
- an EU task force of field-specific regulators for AI/big data
- an EU-level special list of robot rights
- a general fund for all smart autonomous robots
- mandatory consumer protection impact assessment
- regulatory sandboxes
- three-level obligatory impact assessments for new technologies
- the use of anti-trust regulations to break up big tech and appoint regulators
- voluntary/mandatory certification of algorithmic decision systems

Using a pre-defined evaluation strategy, all of these proposals were evaluated. The overall evaluation suggested that many of these options were broad in scope and lacked specific requirements (Rodrigues et al. 2020). They over-focused on well-established issues like bias and discrimination but neglected other human rights concerns, and resource constraints would arise from resource-intensive activities such as the creation of regulatory agencies and the mandating of impact assessments.

Without going into more detail than appropriate for a Springer Brief, what seems clear is that legislation and regulation will play a crucial role in finding ways to ensure that AI promotes human flourishing. A recent review of the media discourse of AI (Ouchchy et al. 2020) shows that regulation is a key topic, even though it is by no means agreed whether and which regulation is desirable.

There is, however, one regulatory option currently being hotly debated that has the potential to significantly affect the future shape of technology use in society, and which I therefore discuss separately in the next section

5.1.3 AI Regulator

The creation of a regulator for AI is one of the regulatory options. It only makes sense to have one if there is something to regulate, i.e. if there is regulation that needs to be overseen and enforced. In light of the multitude of regulatory options outlined in the previous section, one can ask whether there is a need for a specific regulator for AI, given that it is unclear what the regulation will be.

It is again instructive to look at the current EU discussion. The EC's White Paper (European Commission 2020a) treads very carefully in this respect and discusses under the heading of "Governance" a network of national authorities as well as sectoral networks of regulatory authorities. It furthermore proposes that a committee of experts could provide assistance to the EC. This shows a reluctance to create a new institution. The European Parliament's counterproposal (2020a) takes a much stronger position. It renews an earlier call for the designation of a "European Agency for Artificial Intelligence". Article 14 of the proposed regulation suggests the creation of a supervisory authority in each European member state (see, e.g., Datenethikkommission 2019) that would be responsible for enforcing ways of dealing with ethical issues of AI. These national supervisory authorities will have to collaborate closely with one another and with the European Commission, according to the proposal from the European Parliament.

A network of regulators, or even the creation of an entire new set of regulatory bodies, will likely encounter significant opposition. One key matter that needs to be addressed is the exact remit of the regulator. A possible source of confusion is indicated in the titles of the respective policy proposals. Where the EC speaks only of artificial intelligence, the European Parliament speaks of AI, robotics and related technologies. The lack of a clear definition of AI is likely to create problems

A second concern relates to the distribution of existing and potential future responsibilities. The question of the relationship between AI supervisory authorities and existing sectoral regulators is not clear. If, for example, a machine learning system used in the financial sector were to raise concerns about bias and discrimination, it is not clear whether the financial regulator or the AI regulator would be responsible for dealing with the issue.

While the question of creating a regulator or some other governance structure capable of taking on the tasks of a regulator remains open, it is evident that it might be a useful support mechanism to ensure that potential regulation could be enforced. In fact, the possibility of enforcement is one of the main reasons for calls for regulation. It has frequently been remarked that talk of ethics may be nothing but an attempt to keep regulation at bay and thus render any intervention impotent (Nemitz 2018, Hagendorff 2019, Coeckelbergh 2019). It is by no means clear, however, that legislative processes will deliver the mechanisms to successfully address the ethics of AI (Clarke 2019a). It is therefore useful to understand other categories of mitigation measures, and that is why I now turn to the proposals that have been directed at organisations.

5.2 Options at the Organisational Level

Organisations, whether public or private, whether profit-oriented or not, play a central role in the development and deployment of AI. Many of the decisions that influence ethical outcomes are made by organisations. Organisations will also reap many of the benefits of AI, most notably the financial benefits of developing or deploying AI. They are therefore intimately involved in the AI ethics discourse. In this section I distinguish between industry commitments, organisational governance and strategic initiatives that organisations of different types can pursue to address ethical issues of AI.

5.2.1 Industry Commitments

To achieve ethical goals in industry, it is often useful for organisations to join forces, for instance in the formulation of ethical aims (Leisinger 2003). While organisations do not necessarily share goals, benefits or burdens, there are certainly groups of organisations that do have common interests and positions. One action that such organisations can pursue is forming associations to formulate their views and feed them into the broader societal discourse. The most prominent example of such an association of organisations is the Partnership on AI,[1] which includes the internet giants – Google, Apple, Facebook, Amazon, Microsoft – as well as a host of academic and civil society organisations. Other associations such as the Big Data Value Association[2] focus on specific issues or areas, such as big data in Europe.

Industry associations might not have the trust of the public when they represent industrial interests, where these are seen to be in opposition to the broader public good. For instance, the comprehensive Edelman Trust Barometer (Edelman 2020: 23) found that 54% of those surveyed believed that businesses were unfair in that they only catered for the interests of the few rather than serving everybody equally and fairly.

For instance, it seems reasonable to assume that one of the main purposes of the Partnership on AI is to lobby political decision-makers in ways that are conducive to the companies. At the same time, it is reassuring, and maybe a testament to the high visibility of AI ethics, that the pronouncements of the Partnership on AI emphasise ethical issues more heavily than most governmental positions. Most of the members of the Partnership on AI are not-for-profit entities, and its statements very clearly position the purpose of AI in what I have described as AI for human flourishing. The Partnership on AI has a set of tenets published on its website which starts by saying:

> We believe that artificial intelligence technologies hold great promise for raising the quality of people's lives and can be leveraged to help humanity address important global challenges such as climate change, food, inequality, health, and education. (Partnership on AI n.d.)

---

[1] https://www.partnershiponai.org/
[2] http://www.bdva.eu/

Cynics might argue that this is an example of ethics washing (Wagner 2018) with the main purpose of avoiding regulation (Nemitz 2018). However, while there may well be some truth in the charge of ethics washing, the big internet companies are publicly and collectively committing themselves to these noble goals. It is also important to see this in the context of other corporate activities, such as Google's long-standing commitment not to be evil,[3] Facebook's decision to institute an ethics review process (Hoffman 2016) and Microsoft's approach to responsible AI (Microsoft n.d.). Each of these companies has individually been criticised on ethical grounds and will likely continue to be criticised, but for the account of AI ethics in this book it is worth noting that they, at least, as some of the most visible developers and deployers of AI, use a rhetoric that is fully aligned with AI for human flourishing.

This opens up the question: what can companies do if they want to make a serious commitment to ethical AI? I will look at some of the options in the following section on organisational governance.

5.2.2 Organisational Governance

Under this heading I discuss a range of activities undertaken within organisations that can help them deal with various ethical or human rights aspects of AI. Most of these are well established and, in many cases, formalised, often under legal regimes. An example of such an existing governance approach is the corporate governance of information technology which organisations can institute following existing standards (ISO 2008). The ethical issues of AI related to large datasets can, at least to some degree, be addressed through appropriate data governance. Organisational data governance is not necessarily concerned with ethical questions (Tallon 2013, British Academy and Royal Society 2017), but it almost invariably touches on questions of ethical relevance. This is more obvious in some areas than in others. In the health field, for example, where the sensitivity of patient and health data is universally acknowledged, data management and data governance are explicitly seen as ways of ensuring ethical goals (Rosenbaum 2010, OECD 2017). The proximity of ethical concerns and data governance has also led to the development of data governance approaches that are explicitly developed around ethical premises (Fothergill et al. 2019).

Data protection is part of the wider data governance field, the part that focuses on the protection of personal data. Data protection is a legal requirement in most jurisdictions. In the EU, where the GDPR governs data protection activities, data protection is relatively clearly structured, and organisations are aware of their responsibilities. The GDPR has brought in some new practices, such as the appointment of a data

---

[3]There has been a change in the way Google communicates this, but in its current exposition of "Ten things we know to be true", Google still asserts, "You can make money without doing evil" (Google n.d.).

protection officer for organisations, a post whose obligation to promote data protection can take precedence over obligations towards the employer. Data protection in Europe is enforced by data protection authorities: in the UK, for example, by the Information Commissioner's Office (ICO). The national authorities are supported by the European Data Protection Board, which promotes cooperation and consistent application of the law. The European Data Protection Supervisor is an independent EU-level supervisory authority that is part of the board. This is an example of a multi-level governance structure that defines clear responsibilities at the organisational level but extends from the individual employee to national and international regulatory activities. I will come back to this in the recommendations chapter as an example of a type of governance structure that may be appropriate for AI more broadly.

Breaches of data protection are, for many companies, a risk that needs to be managed. Similarly, the broader topic of AI ethics may entail risks, not just in terms of reputation, but also of liability, that organisations can try to address through existing or novel risk management processes. Given that risk assessment and management are well-established processes in most organisations, they may well provide the place to address AI ethics concerns. Clarke (2019b) therefore proposes a focus on these processes in order to establish responsible AI.

A downside of the organisational risk management approach to the ethics of AI is that it focuses on risks to the organisation, not risks to society. For broader societal issues to be addressed, the organisational risk management focus needs to broaden beyond organisational boundaries. As Clarke (2019b) rightly states, this requires the organisation to adopt responsible approaches to AI, which need to be embedded in a supportive organisational culture and business purposes that strengthen the motivation to achieve ethically desirable outcomes.

The EU's AI ethics debate seems to lean heavily towards a risk-based approach. This is perfectly reasonable, in that many AI applications will be harmless and any attempt to regulate them would not only be likely to fail, but also be entirely superfluous. However, there are some AI applications that are high-risk and in need of close scrutiny, and it may be impossible to allow some of those to go forward due to ethical considerations (Krafft et al 2020). This raises the question of whose responsibility it is to assess and manage any risks. The European Parliament (2020b) has suggested that the deployer of an AI system is in control of any risks. The level of risk should determine the liability regime under which damage is dealt with. For the deployer to have clarity on the risk level, the European Parliament has suggested that the EC should hold a list of high-risk AI-systems that require special scrutiny and would be subject to a strict liability regime. In the annex to its draft regulation, the European Parliament lists the following technologies: unmanned aircraft, autonomous vehicles (automation level 4 and 5), autonomous traffic management systems, autonomous robots and autonomous public places cleaning devices (Fig. 5.1).

This approach of focusing on high-risk areas has the advantage of legal clarity for the organisations involved. Its weakness is that it makes assumptions about the risk level that may be difficult to uphold. If risk is determined by the "severity of possible harm or damage, the likelihood that the risk materializes and the manner in which the

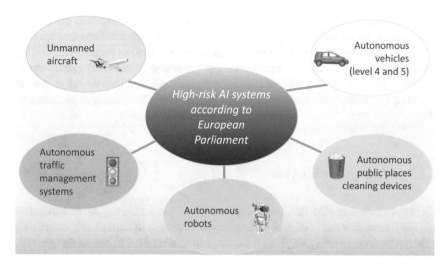

Fig. 5.1 High-risk AI systems according to the European Parliament (2020b)

AI-system is being used" (European Parliament 2020b: art. 3(c)), then it is difficult to see how an abstract list like the one illustrated in Figure 5.1 can determine the risk level. The examples given in the draft regulation are all of systems that carry a risk of physical injury, which is understandable in the context of a liability regulation that is strongly influenced by liability for existing systems, notably from the automotive sector. It is not clear, however, how one would compare the risk of, say, being hit by a falling drone with the risk of being wrongly accused of a crime or the risk of political manipulation of a democratic election.

A risk-based approach nevertheless seems likely to prevail, and there are good reasons for this. The German Datenethikkommission (2019) has proposed a regulation system that may serve as a good example of the risk-based approach. The process of allocating AI systems to these (or similar) risk schemes will be key to the success of a risk-based approach to AI ethics at a societal level, which is a condition for organisations to successfully implement it.

Risk management needs to be based on an understanding of risks, and one aspect of risks is the possible consequences or impacts on society. It is therefore important for organisations aiming to address the ethics of AI proactively to undertake appropriate impact assessments. Some types of impact assessment are already well established, and many organisations are familiar with them. Data protection impact assessments, a development of privacy impact assessments (Clarke 2009, ICO 2009, CNIL 2015), for example, form part of the data protection regime established by the GDPR and are thus implemented widely. Other types of impact assessment cover the environmental impact (Hartley and Wood 2005), the social impact (Becker 2001, Becker and Vanclay 2003), the ethical impact (Wright 2011, CEN-CENELEC 2017) and any impact on human rights (Latonero 2018).

Overall, the various measures that contribute to good organisational governance of AI constitute an important part of good practice that organisations can adopt in order to reduce risks from AI. They may desire to take these measures because they want to do the right thing, but a diligent adoption of good practice can also serve as a defence against liability claims if something goes wrong. This points to the last aspect of organisational responses I want to discuss here, the strategic commitments of an organisation.

5.2.3   Strategic Initiatives

Many companies realise that their responsibilities are wide-ranging and therefore include a commitment to ethical principles and practices in their strategic thinking. This can be done in many ways. The most common term used to denote an organisation's commitment to the greater good is "corporate social responsibility" (CSR) (Garriga and Melé 2004, Blue & Green Tomorrow 2013, Janssen et al. 2015). Libraries have been written about CSR. For the purposes of this book, it suffices to say that CSR is well established as a concept recognised by organisations that may well serve as a starting point for discussing ethical aspects of AI.

One activity often fostered by CSR and arguably of central importance to ensuring adequate coverage of ethical issues in organisations is stakeholder engagement. Stakeholders, following Freeman and Reed (1983), are individuals or groups who are significantly affected by an action or potentially at risk, who thus have a "stake" in it (Donaldson and Dunfee 1999). The term "stakeholder" was coined by Freeman and his collaborators (Freeman and Reed 1983) as a counterpoint to the exclusive focus on shareholders. Stakeholder engagement is now well recognised as a way for organisations to better understand their environment (O'Riordan and Fairbrass 2014). In the ICT world, of which AI forms a part, there can be an affinity between stakeholder engagement and user engagement (Siponen and Vartiainen 2002). Users are understood to be those people who will benefit from a company's products or services, once launched, while stakeholders are those who may experience an impact from the company's work, whether they are users of the product or service or not.

Stakeholder engagement can cover a broad range of activities, and there is little agreement on which methods should be employed to ensure ethically acceptable outcomes. A further and more structured way for organisations to flag their strategic desire to take ethical issues seriously, which may include stakeholder engagement but goes beyond it, is the integration of human rights into organisational strategy and practices.

As a number of the most prominent AI ethics issues are also human rights issues (privacy, equality and non-discrimination), there have been calls for governments, and also private-sector actors, to promote human rights when creating and deploying AI (Access Now Policy Team 2018). The exact nature of the relationship between ethics and human rights is up for debate. While they are not identical, they are at least synergistic (WEF 2018).

Fortunately, the question of the integration of human rights into organisational processes is not entirely new. The UN developed guiding principles for business and human rights that provide help in implementing the UN "protect, respect and remedy" framework (United Nations 2011). While these are generic and do not specifically focus on AI, there are other activities that develop the thinking about AI and human rights further. The Council of Europe has developed principles for the protection of human rights in AI (Commissioner for Human Rights 2019) and more detailed guidance tailored for businesses has been developed by BSR (Allison-Hope and Hodge 2018).

The preceding sections have shown that there are numerous options for organisations to pursue, if they want to address ethical issues of AI. A key question is whether organisations actually realise and implement these options.

5.2.4   Empirical Insights into AI Ethics in Organisations

As part of the empirical case studies undertaken to understand the social reality of AI ethics (see Macnish et al. 2019), respondents were asked about their organisational responses to these issues. It is worth highlighting, however, that from the case studies it became clear that organisations are highly sensitive to some issues, notably those specific issues related to machine learning that are prominently discussed in the media, such as bias, discrimination, privacy and data protection, or data security. Figure 5.2 summarises and categorises the strategies pursued by the organisations investigated.

The organisations researched spent significant efforts on awareness raising and reflection, for example through stakeholder engagement, setting up ethics boards

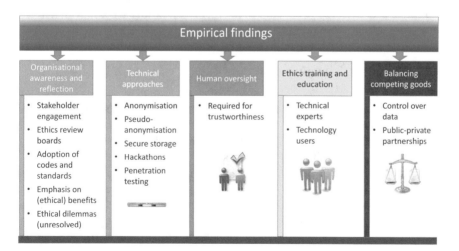

Fig. 5.2   How case study organisations address ethical issues of AI: empirical findings

and working with standards, and they explicitly considered dilemmas and questions on how costs and benefits could be balanced. They particularly employed technical approaches, notably for data security and data protection. There was repeated emphasis on human oversight, and several of the companies offered training and education. In their attempts to balance competing goods, they sometimes sought organisational structures such as public-private partnerships that could help them find shared positions.

The research at the organisational level showed that public and private organisations in Europe take AI ethics very seriously, even though the sample size was not sufficient to make this claim broadly. The organisations engaged with the topic proactively and were considering or already had in place several measures to address ethical challenges. It is noticeable that these measures focused on a subset of the ethical issues described earlier, notably on the specific issues arising from machine learning and in particular those that were already well regulated, such as data protection.

Similarly, the organisations in the sample did not make use of the entire breadth of organisational strategies suggested by the literature. They were not part of any industry associations that aimed to influence the AI ethics environment. And while they probably had organisational risk management or impact assessment structures, these were not highlighted as key to addressing the ethics of AI. Stakeholder engagement was a prominent tool in their inventory. And while they recognised the importance of human rights, they did not make use of formalised methods for the integration of human rights into their processes.

To summarise, one can say that empirical findings from work with organisations suggest that, despite a high level of interest and awareness of AI ethics, there are still numerous options that could be used more widely and there is ample room for development.

This leads to the next point, which is the question of what individuals within and outside organisations can do in order to better understand the ethical issues of AI, and which activities can be undertaken in order to deal with such issues effectively.

5.3 Guidance Mechanisms

The term "guidance mechanisms" is used to describe the plethora of options and support mechanisms that are meant to help individuals and organisations navigate the waters of AI ethics, a very dynamic environment with many actors contributing to the debate and providing tools.

This section presents a brief overview of some of the current activities. It sets out to provide an illustration of some of the options that are available and that complement the policy and organisational level activities. The guidance mechanisms are not independent of policy and organisational options, but often underpin them or result from them and offer ways of implementing them. Some of the guidance mechanisms listed here predate the AI ethics debate but are applicable to it, whereas others have been created in direct response to AI ethics.

The first set of mechanisms consists of guidelines that aim to help users navigate the AI ethics landscape. The most prominent of these from a European perspective were developed by the High Level Expert Group on AI (2019) that was assembled by the European Commission. These guidelines stand out because of their direct link to policymakers, and they are likely to strongly influence European-level legislation on AI. They are by no means the only set of guidelines. Jobin et al. (2019) have identified 84 sets of AI ethics guidelines. In a related study an additional nine sets of guidelines were found (Ryan and Stahl 2020). And there is no doubt that the production of guidelines continues, so that by the time these words are seen by a reader, there will be more.

Jobin et al (2019) do an excellent job of providing an overview of the guidelines landscape and the common themes and threads that pervade it. They also show that there are common themes that cut across them, and are good ways of highlighting key principles and spelling out general expectations. The guidelines have a number of shortcomings, though. They tend to be high-level and therefore not to provide immediately actionable advice. The EU's High Level Expert Group (2020) is therefore developing more applicable tools.

In addition to questions of implementation, there are several more general concerns about guidelines. The large number of these guidelines and their underlying initiatives can cause confusion and ambiguity (Floridi and Cowls 2019). There is a suspicion that they may be dominated by corporate interests (Mittelstadt 2019), a concern that has been prominently voiced by members of the High Level Expert Group (Metzinger 2019). Guidelines can be interpreted as examples of ethics washing or of avoiding legislation (Hagendorff 2019), as noted earlier.

Notwithstanding their disadvantages, ethics guidelines and frameworks are likely to remain a key aspect of the AI ethics debate. Some of them are closely connected with professional bodies and associations, which can help in the implementation phase. Some professional bodies have provided specific guidance on AI and ethics (IEEE 2017, USACM 2017). In addition, they often include AI ethics questions as part of their broader remit. The Association for Computing Machinery (ACM), for example, the largest professional body in computing, has recently refreshed its code of conduct with a view to ensuring that it covers current challenges raised by AI (Brinkman et al. 2017). While professionalism may well have an important role to play in AI ethics, one important obstacle is that in computing, including AI, professionalism is much less well developed than in other areas, such as medicine and law, where professional governance has powers of enforcement that are missing in computing (Mittelstadt 2019).

Professional bodies often contribute to standardisation and, in some cases, are the owners of standards. In the area of AI there are currently several standardisation activities, notably ISO/IEC JTC 1/SC 42,[4] which includes some references to ethical issues. The most prominent standardisation efforts in terms of the ethical aspects

---

[4]https://www.iso.org/committee/6794475.html

of AI is being undertaken by the Institute of Electrical and Electronics Engineers (IEEE) in its P7000 family of standards[5] (Peters et al. 2020).

Standardisation can be linked to certification, something that the IEEE has pioneered with its ethics certification programme for autonomous and intelligent systems.[6] Standards can be made highly influential in various ways. One way is to legally require certification against a standard. This seems to be a key idea currently proposed by the European Commission (2020a) in its AI White Paper. If implemented, it would mean that AI systems of a pre-defined significant risk level would need to undergo certification to ensure ethical issues are appropriately addressed, an idea that appears to have significant support elsewhere (Krafft et al. 2020).

Standardisation can also influence or drive other activities by defining requirements and activities. A well-established example of this is standardisation in information security, where the ISO 27000 series defines best practice. Standardisation can provide technical and organisational guidance on a range of issues. The IEEE P7000 series is a good example. It aims to provide standardisation for specific issues such as privacy (P7002), algorithmic biases (P2003), safety (P7009) and transparency (P7001).

One type of guidance mechanism that standardisation can help with, but that can also draw on other long-standing sources, is development methodologies. This is the topic of IEEE P7000 (model process for addressing ethical concerns during systems design). The idea that ethical issues can and should be considered early on during the development process is now well established, and is an attempt to address the so-called Collingridge (1981) dilemma or the dilemma of control (see box).

> **The Collingridge Dilemma**
>
> Collingridge observed that it is relatively easy to intervene and change the characteristics of a technology early in its life cycle. However, at this point it is difficult to predict its consequences. Later, when the consequences become more visible, it is more difficult to intervene. This is a dilemma for those wanting to address ethical issues during the development process.

The Collingridge dilemma is not confined to AI. In the field of computing it is compounded by the interpretive flexibility and logical malleability of computing technologies, which are clearly features of AI as well. While the uncertainty about future uses of systems remains a fundamental issue that is impossible to resolve, there have been suggestions for how to address it at least to some degree. Many of these suggestions refer to development methodologies, and most go back to some type of *value-sensitive design* (Friedman et al. 2008, van Wynsberghe 2013). The idea behind these methodologies is generally to identify relevant values that should inform the

---

[5] https://ethicsinaction.ieee.org/

[6] https://standards.ieee.org/industry-connections/ecpais.html

development and use of a technology and then engage with relevant stakeholders in discussion on how this can be achieved.

The most prominent example of such a methodology is that of *privacy by design* (ICO 2008, Cavoukian 2009), which the GDPR now mandates under some circumstances as *data protection by design* (Hansen 2016). Attempts have been made to move beyond the specific issue of privacy and its implementation via data protection and to identify broader issues through *ethics by design* (Martin and Makoundou 2017, Beard and Longstaff 2018, Dignum et al. 2018).

Proposals for development methodologies also cover specific steps of the development life cycle, such as systems testing, for example through ethics penetration testing, an idea taken from computer security practice (Berendt 2019), or adversarial testing (WEF 2019). Winfield and Jirotka (2018) suggest transferring the idea of a black box, well known from the aviation industry, to autonomous systems. This would allow tracing of the course of events in the case of an incident, just as an aeroplane's black box helps us understand the cause of an air traffic accident. In addition there are now development methodologies that specifically aim to address the ethics of AI, such as the VCIO (Values, Criteria, Indicators, Observables) model suggested by the AIEI Group (Krafft et al. 2020) or the Virginia Dignum's ART principles for responsible AI (Accountability, Responsibility, Transparency) (Dignum 2019).

In addition, there is a rapidly growing set of tools to address various aspects of AI ethics (Morley et al. 2019). These are published by groups associated with research funders such as the Wellcome Data Lab (Mikhailov 2019), while others originate from non-governmental and civil society organisations, such as Doteveryone and its consequence scanning kit. (TechTransformed 2019). Yet others are based at universities, such as the AI Now Institute, which published an algorithmic impact assessment (Reisman et al. 2018), and yet more come from professional organisations such as the UK Design Council's Double Diamond (Design Council n.d.). Finally, some sets of tools to address AI ethics originate from companies like PWC, which published a practical guide to responsible artificial intelligence (PWC 2019).

In addition to these guidance mechanisms aimed specifically at providing support for dealing with the ethics challenges of AI, there are many further options originating in activities of science and technology research and reflection that can form part of the broader discourse of how to support AI ethics. These include activities such as the anticipation of future technologies and their ethical issues, some of which are closely linked to digital technology (Brey 2012, Markus and Mentzer 2014, Nordmann 2014), but they can also draw on the broader field of future and foresight studies (UNIDO 2005, Sardar 2010). Stakeholder dialogue and public engagement constitute another huge field of activity that will play a central role in AI ethics, drawing on large amounts of prior work to provide many methodologies (Engage2020 n.d.). A final point worth mentioning here is education, which plays a key role in many of the mitigation options. Teaching, training, awareness raising and educating are cornerstones of facilitating a political discourse and reaching policymakers, but also of eliciting a sense of responsibility from AI developers and deployers.

Table 5.1 summarises the mitigation options discussed in this chapter.

Table 5.1 Overview of options to mitigate ethical issues in AI

Policy level		
Policy initiatives	OECD recommendations	
	Open access / data provision	
	Taxation / redistribution	
	Green ICT	
	Personhood for autonomous agents	
Regulation of AI	Current initiatives	EC AI White Paper
		EU Parliament
		National initiatives
	Existing regulations	Intellectual property law
		Competition law
		Data protection
		Human rights legislation
		Liability law
		Regulatory proposals
		Principles of AI regulation
AI regulator	New AI regulator	
	Expand scope of existing regulator	
	Competence centre for AI	
	Network of regulators	
Organisational level		
Industry commitments	Partnership on AI	
	Big data value association	
Organisational governance	Information governance (ISO)	
	Data governance	
	Data protection	
	Risk management (Clarke 2019b)	
	Impact assessment	Data protection impact assessment
		Environmental impact assessment
		Social impact assessment
		Ethics impact assessment
		Human rights impact assessment
	Duty of care / good practice	
Strategic initiatives	Corporate social responsibility	
	Stakeholder engagement	
	Human rights in companies	Council of Europe
		Artificial intelligence: a rights-based blueprint for business
		UN guiding principles
		Limits of human rights
Guidance mechanisms		
AI ethics guidelines and frameworks		
Professionalism		
Standardisation	IEEE P7000 family	
	ISO/IEC JTC 1/SC 42 – Artificial intelligence	
	Certification (IEEE)	
Technical interventions	Transparent / explainable AI	
	Security of AI	

(continued)

Table 5.1 (continued)

Development methodologies	Value-sensitive design
	Privacy by design
	Ethics by design
	Ethics pen testing
	Adversarial testing
	Black box
	VCIO model
	ART (Dignum 2019)
AI ethics tools	Wellcome data lab
	Agile ethics in AI
	Doteveryone's agile consequence scanning
	Responsible Double Diamond "R2D2"
	AI Now's algorithmic impact assessment framework
	Ethical OS (anticipation of future impact)
	PWC responsible AS toolkit
General supporting activities	Anticipation
	Stakeholder dialogue
	Public engagement
	Education / training

The table represents the ways in which AI ethics may be addressed, highlighting the topics mentioned in the text above. It illustrates key options but cannot claim that all strategies are covered, nor that the individual options available for a particular branch are exhaustive. In fact, many of these, for example the AI ethics tools, and the AI ethics frameworks, embrace dozens if not hundreds of alternatives. Highlighting key strands of current debate demonstrates the richness of the field. One final point that adds to the complexity is the set of stakeholders involved, which I will now address.

5.4 AI Ethics Stakeholders

As noted earlier, and following Freeman and Reed (1983), stakeholders are individuals or groups who are significantly affected by an action or potentially at risk. The concept is extensively used in the organisational literature to help organisations identify whom they need to consider when taking decisions or acting (Donaldson and Preston 1995, Gibson 2000).

There are methodologies for stakeholder identification and engagement which allow for a systematic and comprehensive analysis of stakeholders, including specific stakeholder analysis methods for information systems (Pouloudi and Whitley 1997). One challenge with regard to the identification of stakeholders of AI is that, depending on the meaning of the term "AI" used and the extent of the social consequences covered, most if not all human beings, organisations and governmental bodies are stakeholders. In this context the term loses its usefulness, as it no longer helps analysis or allows conclusions to be drawn.

It is nevertheless useful for the purposes of this book to consider AI stakeholders, as a review of stakeholders informs the overall understanding of the AI landscape and provides important support for the use of the ecosystems metaphor to describe AI. I therefore offer a brief overview of key stakeholder groups and categories, indicating their interests or possible actions, which will be referred to later during the discussion of how AI ecosystems can be shaped.

The categorisation I propose is between policy-oriented bodies, other organisations and individuals. These three groups have different roles in shaping, maintaining and interacting within AI ecosystems. Figure 5.2 gives an overview of the three main groups, including examples of the stakeholders who constitute them. The figure takes the form of a Venn diagram in order to indicate that the different groups are not completely separate but overlap considerably. An individual user, for example, may work in a stakeholder organisation and also be part of standardisation and policy development.

The first stakeholder category in the figure relates to policy. Policymakers and institutions that set policies relevant to AI, including research policy and technology policy, but also other relevant policy, such as policies governing liability regimes, have an important role in shaping how ethical and human rights issues concerning AI can be addressed. This includes international organisations such as the UN, the OECD and their subsidiary bodies, such as UNESCO and the International Telecommunication Union.

The European Union is highly influential in shaping policy within the EU member states, and many of its policies are complied with by non-EU policy bodies. International policy is important because it can drive national policy, where legally binding policy in terms of legislation and regulation is typically located. National parliaments and governments thus play a key role in all policy relating to AI ethics. Regulatory bodies that oversee the implementation of regulation also tend to be situated at national level. Further public bodies that are key stakeholders in AI debates are research funding bodies, which can translate policy into funding strategies and implementation requirements. These are often part of public bodies, but they can also be separately funded, as in the case of charitable funders. In Figure 5.3, I have situated ethics bodies in the category of policy. These ethics bodies include the EU's High-Level Expert Group on AI, and also national ethics committees and research ethics committees or institutional review boards, which translate general principles into research practice and oversee detailed implementation at the project level.

The second stakeholder category suggested in Figure 5.3 is that of organisations. This group includes numerous and often very different members. It could easily be broken down further into sub-categories. For the purpose of this book, the members of this second category are categorised by the fact that, as organisations, they have some level of internal structure and temporal continuity and their main purpose is not to develop or implement international or governmental policies.

Key members of this stakeholder category are commercial organisations that play a role in the development, deployment and use of AI. This includes not only companies that develop and deploy AI on a commercial basis, but also users and companies that

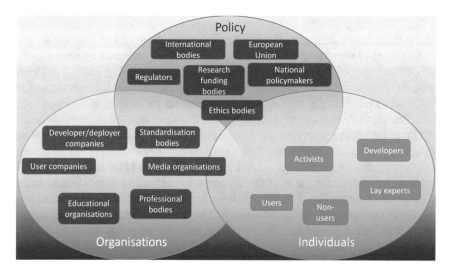

Fig. 5.3 Overview of AI stakeholders

have a special role to play, for example insurance, which facilitates and stabilises liability relationships.

There are numerous organisations that are not in the business of making profits from AI but are involved in the AI value chain, in particular professional bodies, standardisation bodies and educational institutions. These must be included because they have an obvious relationship to some of the mitigation strategies discussed earlier, notably the use of professional standards, the integration of ethical considerations into standards, and the raising of awareness and knowledge through education. Similarly, media organisations play a crucial role in raising awareness of ethical issues and driving public discourse, which in turn may motivate policy development.

The third and final category of stakeholders in this overview is individuals. Policy bodies and organisations are made up of individuals and would cease to exist without individual members. In the category of individuals it is nevertheless important to highlight that there are individuals with characteristics that may not be covered or represented in the other stakeholder groups who still have a legitimate claim to be heard.

Some of these individual stakeholders correspond to organisational stakeholder groups. A developer may be a large profit-driven company, but AI applications can also be developed by a hobby technologist who has the expertise to build novel ideas or applications. Similarly, there are corporate end users of AI, but these tend to have different interests and motivations from individual end users. Activists and lay experts, however, tend to contribute to the public discourse and thereby shape perceptions. But maybe the most important individual stakeholders, because they are most easily overlooked, are those who are characterised by the fact that they do not have an active interest in AI and do not seek to shape it, but are still affected by its existence. These may be individuals who do not have access to AI technologies or facilitating technologies and are therefore excluded from possible benefits, thus raising the problem of digital divides. Another example would be individuals who are subjected to the impact of AI but have no voice, no input and no choice in the matter. Examples of this group would be prisoners whose parole decisions are made using AI or patients whose diagnosis and treatment depend on AI. These are groups that stand to substantially benefit from AI or possibly suffer from it, and thus fulfil the definition of "stakeholder", but often have no way of making their voices heard.

The point of this overview of AI stakeholders has been to demonstrate the complexity of the stakeholder population and indicate the manifold and often contradictory interests that these stakeholders may have. This view of the stakeholder distribution adds to the earlier views of the definition of "AI", the review of ethical issues and mitigation strategies. It shows that there is no simple and straightforward way to drive change and promote flourishing. It is important, for example, to understand that AI can lead to biases and discrimination and that the workings of AI may be non-transparent. But, in order to come to a sound ethical assessment, one needs to understand the detailed working of the technology in its context of use.

In order to say anything useful about how ethical aspects of AI at a more general level can be understood, evaluated and dealt with, it is therefore important to take a different perspective: one that allows us to look at all the various aspects and components, but also allows for a higher-level overview. I therefore suggest that the ethics of AI debate could benefit from a systems-level view, which I introduce in the next chapter.

References

Access Now Policy Team (2018) The Toronto declaration: protecting the right to equality and non-discrimination in machine learning systems. Access Now, Toronto. https://www.access now.org/cms/assets/uploads/2018/08/The-Toronto-Declaration_ENG_08-2018.pdf. Accessed 26 Sept 2020

AI4EU (n.d.) Observatory. https://www.ai4eu.eu/observatory. Accessed 7 Oct 2020

AlgorithmWatch (n.d.) AI ethics guidelines global inventory. https://inventory.algorithmwatch.org/. Accessed 7 Oct 2020

Allison-Hope D, Hodge M (2018) Artificial intelligence: a rights-based blueprint for business. Paper 3: Implementing human rights due diligence. BSR. https://www.bsr.org/reports/BSR-Artificial-Intelligence-A-Rights-Based-Blueprint-for-Business-Paper-03.pdf. Accessed 6 Oct 2020

Beard M, Longstaff S (2018) Ethical by design: principles for good technology. The Ethics Centre, Sydney. https://ethics.org.au/wp-content/uploads/2018/05/The-Ethics-Centre_PRINCI PLES-FOR-GOOD-TECHNOLOGY-29JAN.pdf. Accessed 5 Oct 2020

Becker HA (2001) Social impact assessment. Eur J Oper Res 128:311–321. https://doi.org/10.1016/ S0377-2217(00)00074-6

Becker HA, Vanclay F (eds) (2003) The international handbook of social impact assessment: conceptual and methodological advances. Edward Elgar Publishing, Cheltenham, UK

Berendt B (2019) AI for the common good?! Pitfalls, challenges, and ethics pen-testing. Paladyn J Behav Robot 10:44–65. https://doi.org/10.1515/pjbr-2019-0004

Blue & Green Tomorrow (2013) The guide to corporate social responsibility 2013. http://blueandgr eentomorrow.com/wp-content/uploads/2013/04/BGT-Guide-to-CSR-10MB.pdf. Accessed 5 Oct 2020

Brey PAE (2012) Anticipatory ethics for emerging technologies. NanoEthics 6:1–13. https://doi. org/10.1007/s11569-012-0141-7

Brinkman B, Flick C, Gotterbarn D et al (2017) Listening to professional voices: draft 2 of the ACM code of ethics and professional conduct. Commun ACM 60:105–111. https://doi.org/10. 1145/3072528

British Academy, Royal Society (2017) Data management and use: governance in the 21st century. A joint report by the British Academy and the Royal Society, London. https://royalsociety.org/-/ media/policy/projects/data-governance/data-management-governance.pdf. Accessed 5 Oct 2020

Cavoukian A (2009) Privacy by design: the 7 foundational principles. Information and Privacy Commissioner of Ontario, Ontario. https://www.ipc.on.ca/wp-content/uploads/resources/7found ationalprinciples.pdf. Accessed 6 Oct 2020

CEN-CENELEC (2017) Ethics assessment for research and innovation, part 2: ethical impact assessment framework. CWA 17145-2. European Committee for Standardization, Brussels. http://ftp. cencenelec.eu/EN/ResearchInnovation/CWA/CWA17214502.pdf. Accessed 6 Oct 2020

Clarke R (2009) Privacy impact assessment: its origins and development. Comput Law Secur Rev 25:123–135. https://doi.org/10.1016/j.clsr.2009.02.002

Clarke R (2019a) Regulatory alternatives for AI. Comput Law Secur Rev 35:398–409. https://doi. org/10.1016/j.clsr.2019.04.008

Clarke R (2019b) Principles and business processes for responsible AI. Comput Law Secur Rev 35:410–422. https://doi.org/10.1016/j.clsr.2019.04.007

CNIL (2015) Privacy impact assessment (PIA): good practice. Commission Nationale de l'Informatique et des Libertés, Paris

Coeckelbergh M (2019) Artificial Intelligence: some ethical issues and regulatory challenges. In: Technology and Regulation, pp 31–34. https://doi.org/10.26116/techreg.2019.003

Collingridge D (1981) The social control of technology. Palgrave Macmillan, London

Commissioner for Human Rights (2019) Unboxing artificial intelligence: 10 steps to protect human rights. https://rm.coe.int/unboxing-artificial-intelligence-10-steps-to-protect-human-rights-reco/ 1680946e64. Accessed 6 Oct 2020

Committee on Bioethics (DH-BIO) (2019) Strategic action plan on human rights and technologies in biomedicine (2020–2025). Council of Europe. https://rm.coe.int/strategic-action-plan-final-e/ 16809c3af1. Accessed 6 Oct 2020

Datenethikkommission (2019) Gutachten der Datenethikkommission – Kurzfassung. Datenethikkommission der Bundesregierung, Bundesministerium des Innern, für Bau und Heimat, Berlin. https://www.bmi.bund.de/SharedDocs/downloads/DE/publikationen/themen/it-digitalpolitik/gutachten-datenethikkommission-kurzfassung.pdf?__blob=publicationFile&v=4. Accessed 6 Oct 2020

Design Council (n.d.) What is the framework for innovation? Design council's evolved double diamond. https://www.designcouncil.org.uk/news-opinion/what-framework-innovation-design-councils-evolved-double-diamond. Accessed 18 June 2020

Dignum V (2019) Responsible artificial intelligence: how to develop and use AI in a responsible way. Springer Nature Switzerland AG, Cham, Switzerland

Dignum V, Baldoni M, Baroglio C et al (2018) Ethics by design: necessity or curse? In: Proceedings of the 2018 AAAI/ACM conference on AI, ethics, and society, New Orleans, February 2018. Association for Computing Machinery, New York, pp 60–66

Donaldson T, Dunfee TW (1999) Ties that bind: a social contracts approach to business ethics. Harvard Business Press, Cambridge MA

Donaldson T, Preston LE (1995) The stakeholder theory of the corporation: concepts, evidence, and implications. Acad Manage Rev 20:65–91. https://doi.org/10.2307/258887

Dutton T (2018) An overview of national AI strategies. https://medium.com/politics-ai/an-ove rview-of-national-ai-strategies-2a70ec6edfd. Accessed 7 Oct 2020

Edelman (2020) 2020 Edelman trust barometer global report. https://edl.mn/2NOwltm. Accessed 7 Oct 2020

Engage2020 (n.d.) Action catalogue. http://actioncatalogue.eu/. Accessed 18 June 2020

European Commission (2020a) White Paper on artificial intelligence: a European approach to excellence and trust. European Commission, Brussels. https://ec.europa.eu/info/sites/info/files/commission-white-paper-artificial-intelligence-feb2020_en.pdf. Accessed 22 Sept 2020

European Commission (2020b) Report on the safety and liability implications of artificial intelligence, the internet of things and robotics. European Commission, Brussels. https://ec.europa.eu/info/files/commission-report-safety-and-liability-implications-ai-internet-things-and-robotics_en. Accessed 22 Sept 2020

European Commission (2020c) A European strategy for data. Communication from the Commission to the European Parliament, the Council, the European Economic and Social Committee and the Committee of the Regions. European Commission, Brussels. https://eur-lex.europa.eu/legal-content/EN/TXT/PDF/?uri=CELEX:52020DC0066&from=EN. Accessed 6 Oct 2020

European Commission (2020d) Shaping Europe's digital future. Communication from the Commission to the European Parliament, the Council, the European Economic and Social Committee and the Committee of the Regions. European Commission, Brussels. https://ec.europa.eu/info/sites/info/files/communication-shaping-europes-digital-future-feb2020_en_3.pdf. Accessed 6 Oct 2020

European Commission (n.d.) Knowledge for policy: AI watch https://ec.europa.eu/knowledge4policy/ai-watch_en. Accessed 7 Oct 2020

European Parliament (2020a) Draft report with recommendations to the Commission on a framework of ethical aspects of artificial intelligence, robotics and related technologies. European Parliament, Committee on Legal Affairs. https://www.europarl.europa.eu/doceo/document/JURI-PR-650508_EN.pdf. Accessed 25 Sept 2020

European Parliament (2020b) Draft report with recommendations to the Commission on a civil liability regime for artificial intelligence. European Parliament, Committee on Legal Affairs. https://www.europarl.europa.eu/doceo/document/JURI-PR-650556_EN.pdf. Accessed 6 Oct 2020

Expert Group on Liability and New Technologies (2019) Liability for artificial intelligence and other emerging digital technologies. Publications Office of the European Union, Luxembourg. https://op.europa.eu/en/publication-detail/-/publication/1c5e30be-1197-11ea-8c1f-01aa75ed71a1/language-en/format-PDF. Accessed 23 Sept 2020

Floridi L, Cowls J (2019) A unified framework of five principles for AI in society. Harv Data Sci Rev 1:1. https://doi.org/10.1162/99608f92.8cd550d1

Fothergill BT, Knight W, Stahl BC, Ulnicane I (2019) Responsible data governance of neuroscience big data. Front Neuroinform 13:28. https://doi.org/10.3389/fninf.2019.00028

Freeman RE, Reed DL (1983) Stockholders and stakeholders: a new perspective on corporate governance. Calif Manage Rev 25:88–106

Friedman B, Kahn P, Borning A (2008) Value sensitive design and information systems. In: Himma K, Tavani H (eds) The handbook of information and computer ethics. Wiley Blackwell, Hoboken, NJ, pp 69–102

Garden H, Winickoff DE, Frahm NM, Pfotenhauer S (2019) Responsible innovation in neurotechnology enterprises. OECD Publishing, Paris. https://doi.org/10.1787/9685e4fd-en

Garriga E, Melé D (2004) Corporate social responsibility theories: mapping the territory. J Bus Ethics 53:51–71. https://doi.org/10.1023/B:BUSI.0000039399.90587.34

Gibson K (2000) The moral basis of stakeholder theory. J Bus Ethics 26:245–257. https://doi.org/10.1023/A:1006110106408

Google (n.d.) Ten things we know to be true. https://www.google.com/about/philosophy.html?hl=en_US. Accessed 8 Oct 2020

Hagendorff T (2019) The ethics of AI ethics: an evaluation of guidelines. Minds Mach (Dordr) 30:99–120. https://doi.org/10.1007/s11023-020-09517-8

Hansen M (2016) Data protection by design and by default à la European General Data Protection Regulation. In: Lehmann A, Whitehouse D, Fischer-Hübner S et al (eds) Privacy and identity management: facing up to next steps. Springer International Publishing, Cham, Switzerland, pp 27–38

Hartley N, Wood C (2005) Public participation in environmental impact assessment: implementing the Aarhus convention. Environ Impact Assess Rev 25:319–340. https://doi.org/10.1016/j.eiar.2004.12.002

High-Level Expert Group on Artificial Intelligence (2019) Ethics guidelines for trustworthy AI. European Commission, Brussels. https://ec.europa.eu/newsroom/dae/document.cfm?doc_id=60419. Accessed 25 Sept 2020

High-Level Expert Group on Artificial Intelligence (2020) The assessment list for trustworthy AI (ALTAI). European Commission, Brussels. https://ec.europa.eu/newsroom/dae/document.cfm?doc_id=68342. Accessed 10 Oct 2020

Hoffmann AL (2016) Facebook has a new process for discussing ethics. But is it ethical? The Guardian, 17 June. https://www.theguardian.com/technology/2016/jun/17/facebook-ethics-but-is-it-ethical. Accessed 8 Oct 2020

ICO (2008) Privacy by design. Information Commissioner's Office, Wilmslow, UK. https://web.archive.org/web/20121222044417if_/http://www.ico.gov.uk:80/upload/documents/pdb_report_html/privacy_by_design_report_v2.pdf. Accessed 6 Oct 2020

ICO (2009) Privacy impact assessment handbook, v. 2.0. Information Commissioner's Office, Wilmslow, UK. https://www.huntonprivacyblog.com/wp-content/uploads/sites/28/2013/09/PIA handbookV2.pdf. Accessed 6 Oct 2020

IEEE (2017) The IEEE global initiative on ethics of autonomous and intelligent systems. https://standards.ieee.org/develop/indconn/ec/autonomous_systems.html. Accessed 10 Feb 2018

ISO (2008) BS ISO/IEC 38500:2008: Corporate governance of information technology. British Standards Institute, London

Janssen C, Sen S, Bhattacharya C (2015) Corporate crises in the age of corporate social responsibility. Bus Horiz 58:183–192. https://doi.org/10.1016/j.bushor.2014.11.002

Jobin A, Ienca M, Vayena E (2019) The global landscape of AI ethics guidelines. Nat Mach Intell 1:389–399. https://doi.org/10.1038/s42256-019-0088-2

Krafft T, Hauer M, Fetic L et al (2020) From principles to practice: an interdisciplinary framework to operationalise AI ethics. VDE and Bertelsmann Stiftung. https://www.ai-ethics-impact.org/resource/blob/1961130/c6db9894ee73aefa489d6249f5ee2b9f/aieig—report—download-hb-data.pdf. Accessed 26 Sept 2020

Latonero M (2018) Governing artificial intelligence: upholding human rights & dignity. Data Soc. https://datasociety.net/wp-content/uploads/2018/10/DataSociety_Governing_Artificial_Intelligence_Upholding_Human_Rights.pdf. Accessed 26 Sept 2020

Leisinger KM (2003) Opportunities and risks of the United Nations Global Compact: the Novartis case study. J Corpor Citizensh 11:113–131. http://www.jstor.orgx/stable/jcorpciti.11.113

Macnish K, Ryan M, Gregory A et al (2019) SHERPA deliverable D1.1 Case studies. De Montfort University. https://doi.org/10.21253/DMU.7679690.v3

Markus ML, Mentzer K (2014) Foresight for a responsible future with ICT. Inf Syst Front 1–16. https://doi.org/10.1007/s10796-013-9479-9

Martin CD, Makoundou TT (2017) Taking the high road: ethics by design in AI. ACM Inroads 8:35–37. https://doi.org/10.1145/3148541

Metzinger T (2019) Ethics washing made in Europe. Der Tagesspiegel. https://www.tagesspiegel.de/politik/eu-guidelines-ethics-washing-made-in-europe/24195496.html. Accessed 6 Oct 2020

Microsoft (n.d.) Responsible AI. https://www.microsoft.com/en-us/ai/responsible-ai. Accessed 8 Oct 2020

Mikhailov D (2019) A new method for ethical data science. Wellcome Trust, London. https://welcome.ac.uk/news/new-method-ethical-data-science. Accessed 18 June 2020

Mittelstadt B (2019) Principles alone cannot guarantee ethical AI. Nat Mach Intell 1:501–507. https://doi.org/10.1038/s42256-019-0114-4

Morley J, Floridi L, Kinsey L, Elhalal A (2019) From what to how: an overview of AI ethics tools, methods and research to translate principles into practices. https://arxiv.org/pdf/1905.06876.pdf

Nemitz P (2018) Constitutional democracy and technology in the age of artificial intelligence. Phil Trans R Soc A 376:20180089. https://doi.org/10.1098/rsta.2018.0089

NESTA (n.d.) Mapping AI governance. National Endowment for Science, Technology and the Arts. https://www.nesta.org.uk/data-visualisation-and-interactive/mapping-ai-governance/. Accessed 7 Oct 2020

Nordmann A (2014) Responsible innovation, the art and craft of anticipation. J Respons Innov 1:87–98. https://doi.org/10.1080/23299460.2014.882064

O'Riordan L, Fairbrass J (2014) Managing CSR stakeholder engagement: a new conceptual framework. J Bus Ethics 125:121–145. https://doi.org/10.1007/s10551-013-1913-x

OECD (2017) Recommendation of the OECD Council on health data governance

OECD (2019) Recommendation of the Council on artificial intelligence. OECD/LEGAL/0449. https://legalinstruments.oecd.org/en/instruments/OECD-LEGAL-0449. Accessed 12 Oct 2020

OECD (n.d.) AI policy observatory. Organisation for Economic Co-operation and Development. https://www.oecd.org/going-digital/ai/. Accessed 7 Oct 2020

Ouchchy L, Coin A, Dubljević V (2020) AI in the headlines: the portrayal of the ethical issues of artificial intelligence in the media. AI & Soc. https://doi.org/10.1007/s00146-020-00965-5

Partnership on AI (n.d.) Tenets. https://www.partnershiponai.org/tenets/. Accessed 8 Oct 2020

Peters D, Vold K, Robinson D, Calvo RA (2020) Responsible AI: two frameworks for ethical design practice. IEEE-TTS 1:34–47. https://doi.org/10.1109/TTS.2020.2974991

Pouloudi A, Whitley EA (1997) Stakeholder identification in inter-organizational systems: gaining insights for drug use management systems. Eur J Inf Syst 6:1–14. https://doi.org/10.1057/palgrave.ejis.3000252

PWC (2019) A practical guide to responsible artificial intelligence. https://www.pwc.com/gx/en/issues/data-and-analytics/artificial-intelligence/what-is-responsible-ai/responsible-ai-practical-guide.pdf. Accessed 18 June 2020

Reisman D, Schultz J, Crawford K, Whittaker M (2018) Algorithmic impact assessments: a practical framework for public agency accountability. AI Now Institute, New York. https://ainowinstitute.org/aiareport2018.pdf. Accessed 18 June 2020

Rodrigues R, Panagiotopoulos A, Lundgren B et al (2020) SHERPA deliverable 3.3 Report on regulatory options. https://s3-eu-west-1.amazonaws.com/pstorage-dmu-5536699460/24574814/D3.3RegulatoryoptionsforAI30July2020_final_CLEAN1.pdf. Accessed 6 Oct 2020

Rosenbaum S (2010) Data governance and stewardship: designing data stewardship entities and advancing data access. Health Serv Res 45:1442–1455. https://doi.org/10.1111/j.1475-6773.2010.01140.x

Ryan M, Stahl BC (2020) Artificial intelligence ethics guidelines for developers and users: clarifying their content and normative implications. J Inf Commun Ethics Soc. https://doi.org/10.1108/JICES-12-2019-0138

Sardar Z (2010) The namesake: futures; futures studies; futurology; futuristic; foresight – what's in a name? Futures 42:177–184. https://doi.org/10.1016/j.futures.2009.11.001

Siponen MT, Vartiainen T (2002) Teaching end-user ethics: issues and a solution based on universalizability. Commun Assoc Inf Syst 8:422–443. https://doi.org/10.17705/1CAIS.00829

Stix C (n.d.) Writing. https://www.charlottestix.com/european-union-ai-ecosystem. Accessed 22 June 2020

Tallon PP (2013) Corporate governance of big data: perspectives on value, risk, and cost. Computer 46:32–38. https://doi.ieeecomputersociety.org/10.1109/MC.2013.155

TechTransformed (2019) Consequence scanning. Doteveryone. https://doteveryone.org.uk/download/2786/. Accessed 18 June 2020

UNIDO (2005) UNIDO technology foresight manual, vol 1: organization and methods. United Nations Industrial Development Organization, Vienna. http://www.research.gov.ro/uploads/imported/1226911327TechFor_1_unido.pdf. Accessed 6 Oct 2020

United Nations (2011) Guiding principles on business and human rights: implementing the United Nations "protect, respect and remedy" framework. United Nations Human Rights, Office of the High Commissioner, New York and Geneva. https://www.ohchr.org/documents/publications/guidingprinciplesbusinesshr_en.pdf. Accessed 6 Oct 2020

USACM (2017) Statement on algorithmic transparency and accountability. ACM US Public Policy Council, Washington DC. https://www.acm.org/binaries/content/assets/public-policy/2017_usacm_statement_algorithms.pdf. Accessed 26 Sept 2020

van Wynsberghe A (2013) Designing robots for care: care centered value-sensitive design. Sci Eng Ethics 19:407–433. https://doi.org/10.1007/s11948-011-9343-6

Wagner B (2018) Ethics as an escape from regulation: from ethics-washing to ethics-shopping. In: Bayamlioglu E, Baraliuc I, Janssens LAW, Hildebrandt M (eds) Being profiled: cogitas ergo sum. Amsterdam University Press, Amsterdam, pp 84–90

WEF (2018) White Paper: How to prevent discriminatory outcomes in machine learning. World Economic Forum, Geneva. http://www3.weforum.org/docs/WEF_40065_White_Paper_How_to_Prevent_Discriminatory_Outcomes_in_Machine_Learning.pdf. Accessed 6 Oct 2020

WEF (2019) White Paper: Responsible use of technology. World Economic Forum, Geneva. http://www3.weforum.org/docs/WEF_Responsible_Use_of_Technology.pdf. Accessed 6 Oct 2020

Winfield AF, Jirotka M (2018) Ethical governance is essential to building trust in robotics and AI systems. Philos Trans Royal Soc A 376. https://doi.org/10.1098/rsta.2018.0085

Wright D (2011) A framework for the ethical impact assessment of information technology. Ethics Inf Technol 13:199–226. https://doi.org/10.1007/s10676-010-9242-6

Chapter 6
AI Ecosystems for Human Flourishing: The Background

Abstract This chapter analyses the concept of AI ecosystems with a view to identifying how the ecosystem metaphor can help deal with ethical questions. The first step is to introduce in more detail the concept of ecosystems, drawing specifically on the literature on innovation ecosystems. This allows the identification of characteristics of ecosystems such as their openness, the co-evolution and mutual learning of their members, and the interdependence and complex relationship between those members. These characteristics underlie the challenges that an ethics-driven approach to ecosystems must consider.

Keywords AI ecosystems · Innovation ecosystems · Ethics of ecosystems

Stakeholders in AI are numerous and ethical issues are complex. I propose looking at AI from a systems perspective, specifically employing the idea of an AI ecosystem. This chapter will prepare the groundwork for my later recommendations by giving the necessary background on ecosystems.

6.1 An Ecosystems View of AI

Speaking of AI in terms of ecosystems is by now well established. The EC's White Paper sees AI as composed of an *ecosystem* of excellence and an *ecosystem* of trust (European Commission 2020a). The OECD recommends that national policymakers foster a digital *ecosystem* for AI (OECD 2019: 3). Charlotte Stix (n.d.) has used the term to cover EU AI policy. The UK's Digital Catapult (2020) has used the term to refer to AI ethics, to explore which practical lessons can be learned. The first draft of UNESCO's recommendation on the ethics of artificial intelligence suggests that "Member States should foster the development of, and access to, a digital *ecosystem* for ethical AI" (UNESCO 2020).

To use the ecosystem metaphor productively and examine how it can promote our understanding of AI *and* allow us to deduce recommendations on how such an

© The Author(s) 2021
B. C. Stahl, *Artificial Intelligence for a Better Future*,
SpringerBriefs in Research and Innovation Governance,
https://doi.org/10.1007/978-3-030-69978-9_6

ecosystem could be shaped, we need to look at the history of the use of this metaphor, the purposes for which it has been developed and the limitations that it may have.

6.1.1 AI Innovation Ecosystems

The use of terms such as "innovation ecosystem" is relatively widespread in innovation management and related fields and relates only vaguely to the original use of the term in biology (see box).

Ecosystems

The term ecosystem originally stems from biology. According to National Geographic (Rutledge et al. 2011) "an ecosystem is a geographic area where plants, animals, and other organisms, as well as weather and landscapes, work together to form a bubble of life". Outside of biology, ecosystems are regarded as complex, interconnected networks of individual components ranging from the "U.S. television ecosystem" (Ansari et al. 2016) to "ecosystem service assessments [for] mental health" (Bratman et al. 2019) to conclusions that "1.4 million songs are inspired by ecosystems" (Coscieme 2015). It is a popular concept which suggests that the components of the system are a living organism.

One can distinguish several perspectives that are used in the innovation ecosystems literature. The dominant one is the organisational perspective, where researchers employ the term to better understand how organisations can gain a competitive advantage within a system. Viewing the organisation as part of an innovation ecosystem can provide insights into opportunities for growth (Adner 2006). The ecosystems perspective helps organisations understand that they can shape the ecosystem they are part of, but that the overall innovation is at least partly a function of the surrounding ecosystem (Nylund et al. 2019). Recognising this potential, organisations can use the ecosystems view to develop their strategy (Moore 1993) in general, with a particular focus on their innovation management activities (Ritala and Almpanopoulou 2017). One example of a question that the use of the ecosystem metaphor can help to answer is: how and why do organisations become more or less successful? Moore (1993) uses the example of IBM in the context of ecosystems. IBM was one of the most successful members of the new business community or ecosystem based on personal computers. It dominated this system for a while, but then became less profitable and lost its leadership of the market.

While this functional use of the ecosystem metaphor appears to be the dominant one in the fields of business and organisation studies, it is not the only one. An ecosystems perspective can also be used as a theoretical perspective that allows deeper insights into the behaviour of members of the ecosystem more generally.

The ecosystem lens can equally be employed at the social level, for example by the social sciences to interpret the global economy as a living organism with a view to understanding its workings (Gomes et al. 2018).

From the very beginning of the innovation ecosystems literature (Moore 1993), it has been suggested that society can employ this perspective to provide an environment in which ecosystems can thrive. Another suggestion is that the ecosystems terminology can be used to improve the performance of entire innovation ecosystems (Pombo-Juárez et al. 2017). It has also long been recognised that it may be appropriate for entire ecosystems to perish in the interests of society as a whole (Moore 1993).

But what counts as an innovation ecosystem? I use this term broadly, ignoring the conceptual nuances that distinguish between terms such as "business ecosystem" (Gomes et al. 2018), "digital ecosystem" (Senyo et al. 2019), "digital business ecosystem" and "knowledge ecosystem" (Gomes et al. 2018), and further related ideas such as value chains. These distinctions may be valuable for specific purposes, but for the use of the ecosystems concept to develop normative insights into AI, they are of secondary relevance.

More interesting are the characteristics that the various types of ecosystem display. A key characteristic is that ecosystems are the place where evolution occurs. Darwinian evolution is widely accepted in both natural and social sciences as a theory that explains change (Porra 1999). This renders evolutionary theory attractive to fast-moving fields such as innovation. Moore's (1993) seminal article proposing the application of the ecosystem metaphor to socio-technical systems proposes four stages of development of ecosystems: birth, expansion, leadership and self-renewal or death. The adoption of Darwinian evolutionary theory and the application of evolutionary principles to socio-technical systems is contested and ethically problematic, as I will show below. However, it appears to be highly attractive as a general theory of change.

In addition to explaining change, the ecosystems lens can help explain interdependencies between actors and why they develop together or co-evolve (Ritala and Almpanopoulou 2017). This co-evolution of interdependent actors explains why they have to compete *and* cooperate. Innovation ecosystems, like natural ecosystems, are open systems where new actors can emerge and incumbent ones need to react accordingly. Ecosystems can also be visualised as interdependent and interconnected networks in which mutual learning can occur (Nylund et al. 2019). These networks often have one central node (Gobble 2014), which may be what Gomes et al. (2018) call a "keystone leader", i.e. a dominant organisation. In the case of current innovation ecosystems they are often organised around a technology platform.

These insights that are generated using the ecosystem metaphor imply that it is clear where the boundaries of an ecosystem are. Pombo-Juárez et al. (2017) suggest that ecosystems consist of four layers – individuals, organisations, innovation systems and landscapes – which seems to suggest that it is possible to delineate which landscapes with their inhabitants constitute a particular ecosystem.

Examples of innovation ecosystems can be found at different levels of size and complexity. They can be geographically constrained, as in the case of cities (Nylund et al. 2019) or national innovation ecosystems (Gomes et al. 2018), but they can also be organised by other criteria. Adner (2006) gives the example of industries such as commercial printing, financial services, basic materials and logistics provision.

Of interest to this book is that innovation ecosystems are generally recognised to be subject to intervention and change. These interventions can be catalysed by members of the ecosystem, normally organisations, but also individual or non-organisational collective actors. They can also be triggered by actors who are involved in, support or contribute to the innovation ecosystem, but do not necessarily play a part as members of the ecosystem. For instance, a regional ecosystem may be influenced by a national actor, or an ecosystem of AI may be influenced by non-AI technical developments in, say, quantum computing. Innovation ecosystems are notably different from natural ecosystems in that they have an ability to reflect on their status and think about the future (Pombo-Juárez et al. 2017), with a view to changing and improving the situation.

Figure 6.1 summarises some of the key characteristics of innovation ecosystems that render them an interesting metaphor to describe an ensemble of socio-technical actors, including the AI landscape. The representation in the form of overlapping ellipses symbolises that these characteristics are not independent but influence one another.

It is easy to see why the ecosystem metaphor is applied liberally to AI. There are many different actors and stakeholders involved. These interact in complex ways with consequences that are difficult to predict. They are all mutually dependent, even though the disappearance of any one of them will not necessarily damage the overall system. They co-evolve and try to prosper.

Despite these advantages, there are significant drawbacks to the application of the concept of ecosystems to socio-technical systems. Oh et al. (2016) call it a flawed

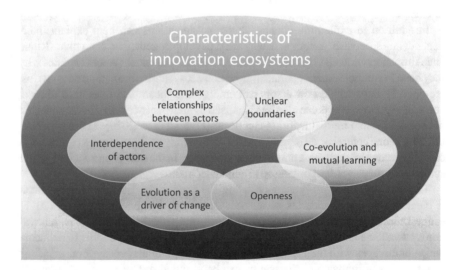

Fig. 6.1 Characteristics of innovation ecosystems

analogy. They point to the fact that, unlike natural ecosystems, innovation ecosystems are not themselves the outcome of evolutionary processes but are intentionally designed. They are concerned that an analogy that does not rest on rigorous conceptual and empirical analysis may preclude more detailed research and policy around innovation. Describing a social system in terms of a natural system furthermore leads to potential conceptual pitfalls. The heavy emphasis on evolutionary processes of selection can lead to an implied technological determinism. This means that the technology in question is seen as an exogenous and autonomous development that is inevitable and forces individuals and organisations to adapt (Grint and Woolgar 1997). Equally, it is problematic that the competitive struggle for survival implied in evolution would then apply not only to organisations but also potentially to cultures where only those who are adapted to the technology survive (Paterson 2007).

There is a well-established link between Darwinism and capitalism (Rauch 1993), with Charles Darwin himself having freely admitted that his theory of evolution was inspired by the classical economists of the 17th and 18th century who focused on the principle of competitive individualism (Priest 2017). Hawkes (2003: 134) therefore goes so far as to call Darwin's theory of evolution a "textbook example of the Marxist theory of ideology in practice", with ideology being a ruling idea of the ruling classes (Shaw 1989).

The innovation ecosystem metaphor can serve ideological purposes by naturalising and thus hiding the fact that the capitalist means of production that these innovation systems are typically based on are the result of historical struggles and political processes that could well have led to other outcomes. To make matters worse, Darwinian ideas developed to describe the natural world have a history of being adopted to legitimate the outcomes of evolution, even where this is arguably inappropriate. This phenomenon is called social Darwinism (Crook 1996). Social Darwinism not only explains social change but can also be used to justify this change. This is a scientifically problematic category mistake which can also have morally outrageous consequences. At its worst, German Nazis adopted social Darwinism in their fantasy that the Aryan "race" was superior to others and needed to preserve its gene pool (Weikart 2013).

The ecosystem metaphor employed here to help explain, understand and guide the field of AI is thus very explicitly not harmless or neutral. The attempt to apply ethical thinking to it raises some further significant issues that I will discuss in more detail below. However, I hope that highlighting these pitfalls will make it possible to avoid them. To emphasise the point, I do not use the concept of ecosystem as a scientific description, but only as a metaphor, that is, a figure of speech and a symbolic representation.

The benefit of this use of the metaphor is that it helps to highlight some key features of AI that contribute to ethical issues *and* affect possible mitigations. I will use the metaphor later to derive some requirements for possible solutions which I hope will allow a more cohesive approach to mitigating ethical risks.

Before we move on to the ethics of ecosystems and the question of how such ecosystems can be shaped, it is important to describe relevant parts of the AI innovation ecosystem. Figure 6.2 provides a systems view of the ecosystem focusing on

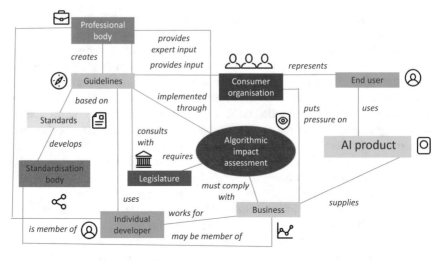

Fig. 6.2 Systems view of stakeholders and activities to implement an algorithmic impact assessment

one particular mitigation strategy, namely the requirement to provide algorithmic impact assessments (see box).

Algorithmic Impact Assessments

Algorithms can be part of systems which make decisions. Algorithmic decision systems (ADS) "rely on the analysis of large amounts of personal data to infer correlations … [and] derive information deemed useful to make decisions". Decisions made by an ADS can be wrong (Oswald et al. 2018). To minimise this risk, algorithmic impact assessments are designed to reduce the risks of bias, discrimination and wrong decision-making.

The introduction of algorithmic impact assessments is one of the numerous proposals for addressing ethical issues in AI that were introduced in Chapter 5. The figure captures the key stakeholders and processes that would be involved in implementing such a proposal. What the example shows is that the characteristics of innovation ecosystems depicted in Figure 6.1, such as non-linearity, interdependence, openness and unclear boundaries, are easily identifiable in practice. This lends credence to my thesis that the innovation ecosystem metaphor is helpful for understanding AI. If this is so, then it is worth taking the next step and thinking about what the metaphor can teach us about intervening in the ecosystem and how we can use it to promote human flourishing.

6.2 Ethics of and in (Eco)Systems

What I have shown so far is that the metaphor of ecosystems is useful because it speaks to an audience of business and political decision-makers. A more general reference to systems theories may help audiences of other types understand the complexity and interdependence of the various human and non-human actors that make up the field of AI. What I have not shown is that the metaphor of ecosystems is useful in understanding AI *ethics* and deriving recommendations for handling AI ethics dilemmas, the ultimate aim of this book. One could object that (eco)systems as phenomena of the natural world can be interpreted as being categorically separate from ethics.

It is a long-established position in philosophical ethics that normative statements (statements that can be expressed using the term "ought") cannot be reduced to or fully deduced from descriptive statements. There is some plausibility to this when applied to ecosystems. Natural ecosystems are the result of evolution and predate humans and human ethical analysis. What individual members of ecosystems do and how ecosystems develop is thus not subject to ethics. If we think of an ecosystem without humans, say a palaeontological ecosystem in the Jurassic period, it is easy to see that an ethical analysis would be difficult and probably meaningless.

However, the relationship between "is" statements and "ought" statements is more complex than this, and numerous voices suggest that descriptive and normative statements at least need to inform one another (Magnani 2007). For instance, "ought implies can" is famously ascribed to Immanuel Kant (Kohl 2015) and links descriptive properties to normative properties. What a human being *can* do is descriptive, and if – as any reasonable person would agree – this needs to inform what *ought* to be done, we have a link. There is also a long tradition of philosophy that derives normative statements from descriptive ones or that derives moral obligations from some sort of being (Floridi 2010).

A different type of argument that seems to preclude the deduction of ethical insights from systems descriptions of reality can be observed in particular streams of systems theory. The epitome of this type of systems thinking is represented by the sociologist Niklas Luhmann (1987), who developed a systems theory based on autopoietic systems. These are systems whose primary purpose is to reproduce themselves. One can interpret biological systems in this way, but Luhmann's focus is on social systems, which follow a particular internal logic and integrate environmental input to maintain the integrity of the system (Introna 1997). The economic system, for example, works on a logic of payments to generate further payments. Raising questions of ethics in such a system either is a mistake of category or will lead to the translation of ethics into payment-related aspects, which is likely to be inappropriate in itself.

The relationship between ethics and ecosystems is thus not straightforward. When deducing ethical statements from a systems description, one should take care to be explicit about the assumptions that support the deduction. The challenge of moving from "is" to "ought" should be addressed. That is a general challenge of this book.

The astute reader will notice that I switched between different concepts of ethics in earlier chapters. I introduced the concept of human flourishing as part of a normative discussion of ethics and concluded that promoting human flourishing ought to be the purpose of AI. In my later discussion of the ethical issues of AI, I took a more descriptive stance, simply accepting as an ethical issue those social facts that people perceive to be ethical issues. This does raise fundamental questions about how we can move from a perception of something as ethically problematic to the normative statement that something should be done, which normally implies an obligation on someone to do something.

Using the ecosystem metaphor to describe ethical issues in AI remains in this tension between "is" and "ought". A description of the ecosystem does not in and of itself provide the basis for normative suggestions on how to deal with it. While I recognise these conceptual issues, I do not view them as insurmountable. Ethical pronouncements do not directly follow from the ecosystem perspective, but ethical pronouncements without a good understanding of the real-life issues, which the ecosystem metaphor provides, would not be useful either.

In public discourse we can observe many examples of normative positions that refer to, for example, natural ecosystems. Saving the environment in general but also safeguarding particular natural ecosystems is widely recognised as not just a possible choice, but a moral obligation. This can be based on a number of normative premises (e.g. one ought to preserve anything living or anything capable of suffering; it is our duty to safeguard an environment fit for human habitation and future generations; one should preserve God's creation; etc.) which are often not explicitly spelled out, but seem to have strong support.

In addition, I tried to provide the normative foundation of the book earlier by drawing on the ancient tradition of human flourishing, which is closely linked to the question of the good life. Our lives take place in natural, social and technical ecosystems, which have a strong bearing on our ability to live well. Drawing on Paul Ricoeur (1999: 256), I suggest that the ethical aim that can motivate the use of technology is the aim of the good life, lived with and for others in just institutions. It is these components that allow for human flourishing, and they motivate and provide the normative underpinnings of the AI-related recommendations that I develop below.

References

Adner R (2006) Match your innovation strategy to your innovation ecosystem. Harv Bus Rev 84:98–107, 148. https://hbr.org/2006/04/match-your-innovation-strategy-to-your-innovation-eco system. Accessed 12 Oct 2020
Ansari S, Garud R, Kumaraswamy A (2016) The disruptor's dilemma: TiVo and the U.S. television ecosystem. Strateg Manag J 37:1829–1853. https://doi.org/10.1002/smj.2442
Bratman GN, Anderson CB, Berman MG et al (2019) Nature and mental health: an ecosystem service perspective. Sci Adv 5. https://doi.org/10.1126/sciadv.aax0903
Castelluccia C, Le Métayer D (2019) Understanding algorithmic decision-making: opportunities and challenges. Scientific Foresight Unit, European Parliamentary Research Service, Brussels.

https://www.europarl.europa.eu/RegData/etudes/STUD/2019/624261/EPRS_STU(2019)624
261_EN.pdf. Accessed 12 Oct 2020

Coscieme L (2015) Cultural ecosystem services: the inspirational value of ecosystems in popular
music. Ecosyst Serv 16:121–124. https://doi.org/10.1016/j.ecoser.2015.10.024

Crook P (1996) Social Darwinism: the concept. Hist Eur Ideas 22(4):261–274. https://doi.org/10.
1016/S0191-6599(96)00005-8

Digital Catapult (2020) Lessons in practical AI ethics: taking the UK's AI ecosystem from "what" to
"how". Digital Catapult, London. https://connectedeverythingmedia.files.wordpress.com/2020/
05/20200430_dc_143_ethicspaper.pdf. Accessed 12 Oct 2020

European Commission (2020a) White paper on artificial intelligence: a European approach to
excellence and trust. European Commission, Brussels. https://ec.europa.eu/info/sites/info/files/
commission-white-paper-artificial-intelligence-feb2020_en.pdf. Accessed 22 Sept 2020

European Commission (2020b) Shaping Europe's digital future. Communication from the Commission to the European Parliament, the Council, the European Economic and Social Committee and
the Committee of the Regions. European Commission, Brussels. https://ec.europa.eu/info/sites/
info/files/communication-shaping-europes-digital-future-feb2020_en_3.pdf. Accessed 6 Oct
2020

Floridi L (2010) Information ethics. In: Floridi L (ed) The Cambridge handbook of information and
computer ethics. Cambridge University Press, UK, pp 77–97

Gobble MM (2014) Charting the innovation ecosystem. Res Technol Manage 57:55–59. https://
www.tandfonline.com/doi/abs/10.5437/08956308X5704005

Gomes LA de V, Facin ALF, Salerno MS, Ikenami RK (2018) Unpacking the innovation ecosystem
construct: evolution, gaps and trends. Technol Forecast Soc Change 136:30–48. https://doi.org/
10.1016/j.techfore.2016.11.009

Grint K, Woolgar S (1997) The machine at work: technology, work and organization. Polity Press,
Cambridge

Hawkes D (2003) Ideology, 2nd edn. Routledge, London

Introna LD (1997) Management, information and power: a narrative of the involved manager.
Palgrave Macmillan, London

Kohl M (2015) Kant and "ought implies can". Philos Q 65:690–710. https://doi.org/10.1093/pq/
pqv044

Luhmann N (1987) Soziale Systeme: Grundriß einer allgemeinen Theorie, 1st edn. Suhrkamp
Verlag, Frankfurt am Main

Magnani L (2007) Morality in a technological world: knowledge as duty. Cambridge University
Press, Cambridge, UK

Moore JF (1993) Predators and prey: a new ecology of competition. Harv Bus Rev 71:75–86. https://
hbr.org/1993/05/predators-and-prey-a-new-ecology-of-competition. Accessed 12 Oct 2020

Nylund PA, Ferras-Hernandez X, Brem A (2019) Strategies for activating innovation ecosystems:
introduction of a taxonomy. IEEE Eng Manage Rev 47:60–66. https://doi.org/10.1109/EMR.
2019.2931696

OECD (2019) Recommendation of the council on artificial intelligence. OECD/LEGAL/0449.
https://legalinstruments.oecd.org/en/instruments/OECD-LEGAL-0449. Accessed 12 Oct 2020

Oh D-S, Phillips F, Park S, Lee E (2016) Innovation ecosystems: a critical examination. Technovation
54:1–6. https://doi.org/10.1016/j.technovation.2016.02.004

Oswald M, Grace J, Urwin S, Barnes G (2018) Algorithmic risk assessment policing models: lessons
from the Durham HART model and 'experimental' proportionality. Inf Commun Technol Law
27:223–250. https://doi.org/10.1080/13600834.2018.1458455

Paterson B (2007) We cannot eat data: the need for computer ethics to address the cultural and
ecological impacts of computing. In: Hongladarom S, Ess C (eds) Information technology ethics:
cultural perspectives. Idea Group Reference, Hershey PA, pp 153–168

Pombo-Juárez L, Könnölä T, Miles I et al (2017) Wiring up multiple layers of innovation ecosystems:
contemplations from Personal Health Systems Foresight. Technol Forecast Soc Change 115:278–
288. https://doi.org/10.1016/j.techfore.2016.04.018

Porra J (1999) Colonial systems. Inf Syst Res 10:38–69. https://doi.org/10.1287/isre.10.1.38. Accessed 12 Oct. 2020

Priest G (2017) Charles Darwin's theory of moral sentiments: what Darwin's ethics really Owes to Adam Smith. J Hist Ideas 78(4):571–593. https://doi.org/10.1353/jhi.2017.0032. Accessed 12 Oct 2020

Rauch J (1993) Kindly inquisitors: the new attacks on free thought. University of Chicago Press, Chicago IL

Ricoeur P (1999) Lectures, vol 1: autour du politique. Points, Paris

Ritala P, Almpanopoulou A (2017) In defense of 'eco' in innovation ecosystem. Technovation 60:39–42. https://doi.org/10.1016/j.technovation.2017.01.004. Accessed 12 Oct 2020

Rutledge K, Ramroop T, Boudreau D (2011) Ecosystem. National Geographic. https://www.nationalgeographic.org/encyclopedia/ecosystem/. Accessed 12 Oct 2020

Senyo PK, Liu K, Effah J (2019) Digital business ecosystem: literature review and a framework for future research. Int J Inf Manage 47:52–64. https://doi.org/10.1016/j.ijinfomgt.2019.01.002

Shaw W (1989) Ruling ideas. Can. J Philos 15(Suppl 1):425–448. https://doi.org/10.1080/00455091.1989.10716806

Stix C (n.d.) Writing. https://www.charlottestix.com/european-union-ai-ecosystem. Accessed 22 June 2020

UNESCO (2020) First draft text of the recommendation on the ethics of artificial intelligence. Ad hoc expert group (AHEG) for the preparation of a draft text, UNESCO, Paris. https://unesdoc.unesco.org/ark:/48223/pf0000373434. Accessed 12 Oct 2020

Weikart R (2013) The role of Darwinism in Nazi racial thought. Ger Stud Rev 36:537–556. https://doi.org/10.1353/gsr.2013.0106

Chapter 7
AI Ecosystems for Human Flourishing: The Recommendations

Abstract This chapter develops the conclusions that can be drawn from the application of the ecosystem metaphor to AI. It highlights the challenges that arise for the ethical governance of AI ecosystems. These provide the basis for the definition of requirements that successful governance interventions have to fulfil. Three main requirements become apparent: the need for a clear delimitation of the boundaries of the ecosystem in question, the provision and maintenance of knowledge and capacities within the ecosystem, and the need for adaptable, flexible and careful governance structures that are capable of reacting to environmental changes. Based on these requirements, the chapter then spells out some recommendations for interventions that are likely to be able to shape AI ecosystems in ways that are conducive to human flourishing.

Keywords Ethical governance of AI · Requirements for ethical AI · Excellence and flourishing · Stakeholder engagement · Regulation and enforcement

There are good reasons for thinking of AI in terms of ecosystems, as shown in Chapter 5. There are also good reasons for thinking of AI *ethics* in terms of ecosystems, as shown in Chapter 6. What remains is the task of translating insights into practical interventions that render the AI ecosystem conducive to human flourishing while taking into account the conceptual and empirical findings presented in Chapters 3 and 4. That is the task undertaken in this chapter.

7.1 Challenges of Ethical Governance of the AI Ecosystem

Let us start with insights gained from the empirical research described earlier. As outlined in Chapter 2, are competing interpretations of the concept of AI and varying views on why and how the technologies that are grouped under this label should be used. Any position taken on these concepts strongly influences the types of

© The Author(s) 2021   91
B. C. Stahl, *Artificial Intelligence for a Better Future*,
SpringerBriefs in Research and Innovation Governance,
https://doi.org/10.1007/978-3-030-69978-9_7

ethical issues that are associated with AI. For instance, machine learning has particular sets of characteristics that raise particular concerns, which are different from those raised by a wider understanding of AI as a socio-technical system with society-wide implications. Artificial general intelligence raises another set of concerns again. The multiplicity of concepts, issues, actions and actors is the motivation behind the choice of the ecosystem metaphor to describe the AI system.

What we can learn from this way of looking at AI is that any intervention at the level of the ecosystem must remain sensitive to this complexity. It must incorporate different understandings of the concepts involved, and take into account the role of and impact on the various stakeholders and the interplay between stakeholders, issues and interventions.

The problem is not only that there are many different issues, actors and responses. The ecosystem metaphor highlights the fact that the relationships between the constituent components of the system are often highly non-linear. This means that we can rarely expect to find simple cause-effect relationships. An intervention at some point of the ecosystem can have unexpected consequences that may have the opposite effect to that intended. This is a well-described phenomenon in natural ecosystems (Tenner 1997) that can be observed in similar ways in socio-technical systems, including AI ecosystems. These systems-related effects contribute to the general problem of unintended consequences.

The idea of intervening in an AI ecosystem in order to promote human flourishing is furthermore complicated by the often unclear and shifting boundaries of ecosystems. The boundary of an ecosystem is at least partly determined by the observer who is taking an interest in the system. Natural ecosystems can provide a good example. We could look at the entire earth as an ecosystem, but this can also be broken down into sub-systems, for example by geographical boundaries, which can again be broken down further, for instance by only looking at the habitat of one species. The value of a particular definition of a system with specified borders depends on what the observer who draws the boundaries wants to achieve.

Similarly, the AI ecosystem is not just one system but a system of systems of systems. For instance, we could look at the *global* AI ecosystem. There are some aspects of AI that are indeed global, notably the principles, techniques and technologies, and some of the dominant companies that have a global reach and presence. At the same time one can distinguish regional differences, e.g. between the USA, Europe and China, which could be described as separate ecosystems (see the discussion of different purposes of AI use in Section 3.3). The differentiation by geography and jurisdiction could go further, with, for example, the German system being different from the Spanish one, as shown by Kriechgaum et al. (2018) using the example of the innovation ecosystems surrounding photovoltaics. One can also differentiate further between AI ecosystems for different sectors or areas of application, such as autonomous transport, education, production and agriculture.

All AI ecosystems are embedded in environments which partly shape them but in turn are shaped by them. This throws up further challenges of governance, as any intervention tries to hit moving, interconnected targets. These environments cover technical, policy, economic, legal, social, ethical and other aspects that closely interact with AI and are very influential in the way ethical and related issues materialise, are perceived and can be addressed. They raise problems because their rate of change is likely to be different from that of the AI ecosystem.

As an example, let us look at the legal system, and more specifically at legal liability. Legal liability rules for AI are likely to have a significant impact on the way societies deal with AI. It is therefore not surprising that there are several reviews and recommendations at the European level alone which reflect on the applicability and possible development of liability legislation to render it suitable for AI (Expert Group on Liability and New Technologies 2019, European Commission 2020a, European Parliament 2020a). Liability legislation could therefore be considered a component of the AI ecosystem. At the same time, apart from existing black-letter law, there are also common-law and other legal practices and experiences. Legal professionals with expertise in liability do not necessarily have expertise in AI. Hence, there are different expectations from different fields of application of AI that will conceptualise liability differently. The interaction between the AI ecosystem (with its sub-systems) and the legal liability regime is likely to be complex.

Similar constellations are likely to be relevant to other social or technical systems. Let us take the technical system as an example: AI relies on existing and future ICT infrastructure, such as networking, computing and storage capacity. Progress in these areas has been a crucial factor in the success of machine learning. The availability of appropriate energy sources is a technical challenge but increasingly also a social, political and environmental one, due to the ever-increasing power consumption of AI systems and the potential interference with sustainability goals (Knight 2020). The AI ecosystem is thus crucially dependent on the technical infrastructure, and drives and shapes its development. But decisions about the technical infrastructure are not necessarily taken by members of the AI ecosystem and can therefore appear to be part of the external environment. Evaluation of the state of the ecosystem and perceptions of its progress, potential and ability to change will therefore depend heavily on where exactly the boundary is drawn.

A further key challenge for the ethical governance of AI ecosystems is the concept of ethics. In Chapter 2 I proposed the concept of human flourishing as the concept to guide the understanding of ethics in AI. "Flourishing" is a well-established term strongly linked to the ancient tradition of virtue ethics; it is an inclusive term that is open to figures of thought from other philosophical traditions, such as utility, duty and care. At the same time this openness can be problematic because it is difficult to determine when and how exactly flourishing has been achieved (see box).

Determining Flourishing

One approach that aims for human flourishing, and simultaneously tries to provide concrete guidance on how to measure conditions for flourishing, was developed by Martha Nussbaum: the capabilities approach (Nussbaum 2000, Buckingham n.d.). The ten capabilities required for flourishing, according to Nussbaum, are life; bodily health; bodily integrity; senses, imagination and thought; emotions; practical reason; affiliation; other species (e.g. connection to animals and nature); play; and control over one's environment (Nussbaum 2000: 78–80). Economists have taken Nussbaum's work and assessed whether capabilities and related interventions can be reliably measured (Anand et al 2008). Their conclusion is that economic models focused on Nussbaum's capabilities *can* measure and address some inhibitors of human flourishing, but not all (ibid. 303) due to the fact that capabilities have multiple dimensions (ibid. 302).

We therefore cannot assume that there is an a priori way of determining whether people are flourishing, so we need to concede that this is at least partly an empirical matter which is also subject to change over time. People's moral perceptions and positions change, while AI ecosystems are realised within the shifting boundaries of ethical preferences. At present this may best be illustrated using the different privacy and data protection regimes in different parts of the world, which arguably reflect different social preferences and give rise to interesting debates about what, if anything, is universal and should be applied across geographical and other boundaries. For instance, the right to privacy is recognised as a human right in the European Convention on Human Rights, which provides a strong basis for data protection as a crucial component of safeguarding informational privacy. In the EU data protection is regulated through the General Data Protection Regulation, which provides detailed guidance and requires certain activities and procedures, such as the need to have a legal basis for the processing of data and requirements to undertake data protection impact assessments and appoint data protection officers. The European emphasis on data protection is likely to strongly influence how AI will be regulated (EDPS 2020). In other parts of the world privacy and data protection have different roles and relevance. While data protection legislation exists in many jurisdictions, its extent and enforcement varies. In China, for example, privacy laws protect citizens' data from abuse by third parties, but they do not cover governmental data access and use (Gal 2020).

The concept of human flourishing has some universal claims, notably that humans strive for happiness and that achieving this is an ethically justified aim that societies and the technologies they employ ought to support. But how exactly this is achieved and how we can know whether it has been achieved remain open questions. And this openness is not just a historically contingent fact, but part of the question itself. It is not a question that one can expect to answer and close, but one that needs ongoing reflection and discussion, as particular answers differ and vary over time.

This also implies another major challenge to the ethical governance of AI ecosystems, namely the inevitable existence of ethical disagreement and value conflicts. As part of the process of reflecting on and promoting human flourishing, people will come into conflict. Conflicts may be local, for example where scarce resources must be allocated to satisfy competing demands. A typical example would be the use of water to keep golf courses green versus other uses (Scott et al. 2018). But they can also be more fundamental, where ethically well-justified positions come into conflict and resolutions of such conflicts are not obvious and straightforward. An example might be the controversy over mobile tracking and tracing apps during the COVID-19 crisis, in which competing demands from privacy campaigners and public health experts have led to a number of controversies over how such technologies could and should be used to limit the spread of the disease (Klar and Lanzerath 2020). This is also a good example of the problems in drawing a boundary around a socio-technical innovation ecosystem in terms of jurisdictions, data, technical platform etc.

The final challenge to the successful ethical governance of AI ecosystems is the uncertainty of all aspects of the ecosystems themselves and of their environments, be they technical, social or ethical. Technical uncertainty may be the most visible example, with AI-related technical developments happening at a rapid rate, which renders the value of trying to predict the next step exceedingly limited. This is partly a function of the technology itself, but partly also a function of the growing realisation of the potential applications of such technologies. The application of current machine learning technologies may lead to radical changes in coming years even without any further technical progress, simply because actors are beginning to understand what these technologies can do and to apply them to new problems in novel ways.

But the uncertainty of the future is not just linked to technical artefacts. It is equally important in terms of social structures and ethical preferences. Societies are always dynamic, and this can play out in ways that affect technological ecosystems in unpredictable ways. Again, the COVID-19 pandemic can serve as an illustration of the sometimes rapid change of social systems. Widespread working from home may be supported by AI technologies to the benefit of employees, but it can also offer new modes of surveillance and exploitation of workers (Harwell 2020). As another example, the heightened awareness of racism that has arisen in the context of the Black Lives Matter movement has put an even stronger spotlight on bias and discrimination, already a well-discussed topic of AI ethics. While the potential of AI to lead to bias and discrimination is well established, it has also been remarked that it may turn out to be a useful tool in identifying existing human biases and thereby overcoming them (Stone et al. 2016). It is impossible to predict which social change will raise the next set of challenges and how the interaction between the AI ecosystem and other parts of our social and technical environment will develop.

Figure 7.1 summarises the points discussed in this section, in which I have indicated that most of the key challenges are not specific to AI. Some of them arise from the systemic nature of the socio-technical innovation ecosystem. Some of them are related to fundamental aspects of the social, technical and natural world we live in. AI-specific issues that are linked to the characteristics of the underlying technologies and their impact on the world form only a sub-set of these challenges. This indicates that the governance of AI ecosystems is best understood as a part of the governance of digital technologies, which, in turn, is a sub-set of technology governance overall.

Having now explored the challenges that any attempt to govern AI ecosystems to support human flourishing is likely to face, we can move on to the next step, an exploration of what any intervention would need to cover, in order to address the challenges discussed here.

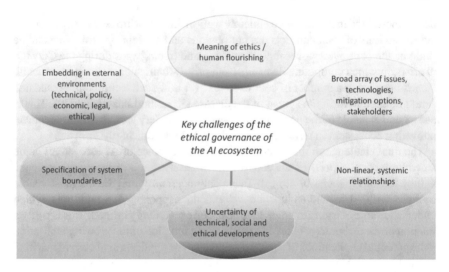

Fig. 7.1 Key challenges of ethical governance of AI ecosystems

7.2   Requirements for Shaping the AI Ecosystem

Interventions designed to address the ethical challenges of AI can be developed in a multitude of ways. The way chosen here is to situate interventions within the AI ecosystem metaphor. A number of requirements for this to work are outlined below.

7.2.1   Clear Boundary of the Ecosystem

The first requirement highlights the importance of drawing clear boundaries around the ecosystem that is to be targeted. This refers to the geographical, technical, cultural and other boundaries that determine what is included in the ecosystem. Boundaries of an ecosystem, as indicated earlier, are not so much natural phenomena as the result of human decisions.

In some cases, these boundaries may seem obvious to the actors involved. In the case of the European Union's discussion of AI, one implicit assumption is that any intervention that the EU undertakes will be at the European level and fall within the EU's jurisdiction. This of course makes perfect sense for legal institutions that work in a defined jurisdiction and therefore intervene within the boundaries of that jurisdiction. The same is true for national regulatory interventions, which are normally aimed at the members of the ecosystem that are active within a nation's borders.

However, it is also clear that the application of jurisdictional boundaries to AI ecosystems is not necessarily the most promising approach. AI principles, the underlying science and algorithms are not locally confined. Many of the key companies in the field are global. The potential mismatch between regional regulation and global technology is not new and not confined to AI (Xu et al. 2004). It is worth being aware of and explicit about it, however, to ensure that the expectations levelled at interventions are realistic.

A similar challenge to the setting of clear boundaries around relevant ecosystems is the terminology describing the underlying technology. As I showed in Chapter 2, the challenge of defining AI means that it is difficult to determine which ethical issues, or which possible mitigations, are relevant. When describing the AI ecosystem, do we refer to AI in a narrow sense, notably to questions of machine learning, neural network, deep learning etc.? Or do we include other aspects of AI, such as fuzzy logic and expert systems? Or do we include the broader socio-technical systems that may or may not embrace narrow AI somewhere along the value chain?

Focusing on AI in a narrow sense has the advantage that technologies are more easily described. The characteristics that give rise to concerns can be identified and a set of ethical issues can be determined. In many cases, the narrower focus might make it easier to find a resolution. A good example of this would be biases that machine learning algorithms pick up from existing datasets which lead to unfair discrimination. This is by now a well-recognised problem and much work is being undertaken to find ways of addressing it (Holzinger et al. 2017).

While such a narrow view of the technologies that constitute the AI ecosystem is thus suitable for resolving particular issues, it is arguably not helpful if one seeks to arrive at a more comprehensive approach that covers the breadth of the current AI ethics discourse. Most of the issues that arise from living in a digitally enabled society go beyond specific technologies and easily identifiable causal chains. While many concerns about fairness, the distribution of opportunities and burdens, employment etc. are related to narrow AI, they typically go beyond the immediate technology used.

The struggle with the question of how to draw the ecosystem boundary can be observed at the European level. For instance, the European Commision's White Paper on Artificial Intelligence (2020c) speaks only of AI and seems to focus on machine learning, but it also refers to other technologies such as quantum computing. This means that no boundary was determined: AI was not defined or set apart from other technologies. In a different report on the safety and liability implications of AI (European Commission 2020a), the technical scope was already broadened in the title to include "the internet of things and robotics". This means that there is a lack of agreement on the exact scope and delimitations of the term even within a single policymaking body such as the European Commission. In its policy outline for Europe's digital future (European Commission 2020b), the authors use the concept of "deep tech", which includes supercomputing, quantum technologies, blockchain and secure pan-European cloud capacities.

This shows the difficulty of clearly delimiting which technologies are of relevance in any given debate or for given reports. One could instead use the term "smart information systems" (SIS), defining SIS as those socio-technical systems that have as part of their core capacities narrow AI and big data analytics but also include other technologies to collect and process data and interact with the external environment.

(Stahl and Wright 2018). This is a useful term, but in the public and policy discourse the term "AI" has become dominant. For the purposes of the delimitation of the ecosystem it is nevertheless important to make clear which technologies are covered and which, by implication, are not.

A third aspect of drawing clear boundaries for the ecosystem, in addition to geographical boundaries and technical terminology, involves its normative aspects. These start with the normative assumptions behind the definition of the ecosystem. Decisions on the geographical, technical and social boundaries of an ecosystem are based on underlying assumptions and values that must be made explicit. What is the observer of the ecosystem who uses the ecosystem metaphor trying to achieve? In Chapter 4 I suggested that there are different purposes that drive the development and use of AI (economic efficiency, social control, human flourishing) and that the delimitation of the ecosystem should include a clarification of which of these (or maybe other) purposes motivate the description of the ecosystem.

If the purpose of using the ecosystem metaphor is to find ways of promoting human flourishing, then this should not only be explicit, but also come with a health warning. Ethical questions are not subject to straightforward resolutions. Anyone explicitly attempting to promote ethics would be well advised to proactively engage in expectation management. Promising to solve the ethics of AI is unlikely to be successful on all counts and may therefore result in disappointment and disillusionment. It might therefore be more fruitful to focus on specific indicators of how human flourishing can be promoted. This may be achieved, for example, by focusing on how some or all the fundamental human rights could be strengthened in an AI ecosystem or by referring to how the AI ecosystem would promote the achievement of the UN's Sustainable Development Goals (SDGs). As such the boundary setting implied in declaring that human flourishing is the ethics focus of the AI ecosystem is narrowed down to specific goals related to promoting human rights.

Defining and delimiting the AI ecosystem in terms of geographical, jurisdictional, cultural or other boundaries, and clarifying the technologies to be covered and the normative aims that are to be achieved, constitute an important first step for a successful intervention in such an AI ecosystem. But on its own this step cannot make a difference.

The next question, then, is: what is required to shape this ecosystem to support human flourishing?

7.2.2 Knowledge and Capacity

One of the characteristic features of innovation ecosystems is that the members of the system not only compete *and* cooperate, but also co-evolve and learn from one another. The existence and availability of knowledge are key factors distinguishing different ecosystems. Knowledge also plays a key role in understanding ethical questions and ways of addressing them. The range and quality of knowledge within an ecosystem are therefore key factors affecting the ability to understand and address ethical concerns.

Shaping an AI ecosystem in a way that promotes human flourishing requires and builds on knowledge. This claim is uncontentious for the technical and economic side of AI ecosystems. The hotbeds of current AI development, notably key geographical areas such as Silicon Valley, are characterised by high levels of available technical talent and knowledge concentrated in high-profile universities and companies. Similarly, building up this technical knowledge base is a key component of most national AI strategies related to particular economies. Such knowledge includes technical knowledge in the narrow sense, but also the knowledge of business processes, finance options and so on that is required for AI organisations to operate successfully.

The same is true for the wider non-technical or normative knowledge that shapes the AI ecosystem. This knowledge covers much of what I have discussed in earlier chapters, such as various concepts of ethics, the ethical issues that are typically associated with AI and the various mitigation options that have been proposed and are being discussed. An ecosystem can only be shaped to promote human flourishing when significant knowledge of ethical issues and potential solutions to ethical challenges is available.

In addition, there is a need for the procedural knowledge that is required to address ethical issues. This is knowledge of how to organise the processes that are required to deal with ethical questions. For instance, if discrimination possibilities are to be reduced to increase human flourishing, the governance options to achieve this need to be known, from legal solutions to technical solutions.

Procedural knowledge should also cover other aspects of reflecting on and evaluating science and technology. Drawing on the discourse on responsible research and innovation, one can identify some important processes that may contribute to ensuring that ethical issues can be recognised and dealt with. These include anticipation, engagement, reflexivity and responsiveness (Stilgoe et al. 2013).

Processes of anticipation are required for a structured way of thinking about possible future states that will inform the way in which we act today and prepare for the future. These processes should not be misinterpreted as simple predictions that try to guess what the future will look like. Accurate predictions are notoriously difficult, if not impossible, and the literature is littered with predictions that have turned out to be wrong and in hindsight tend to look ridiculous, such as the prediction by the president of IBM in 1943 that there would be a world market for maybe five computers, or the statement by the chairman of Digital Equipment Corporation in 1977 that there was no reason for anyone to want a computer in their home (Himanen 2001: 187). Processes of anticipation are based on the recognition of the impossibility of prediction. They nevertheless aim to explore possible futures, to help societies decide which actions to take today (Cuhls 2003). There are well-established discourses and academic disciplines that have developed methods for future and foresight studies (Sardar 2010), some of which explicitly focus on the ethical issues of emerging technologies (Brey 2011, Floridi and Strait 2020). For instance, Flick et al. (2020) explore a wide range of resources, including academic publications, but also social media discussions, to identify expected technical developments in the field of ICT for ageing and ethical concerns that may arise from these. This type of work opens up spaces of possibilities without committing itself to one particular outcome. It is useful in raising awareness and sensitivity to both technical and social or ethical developments and therefore offers the likelihood that these can be shaped appropriately. This type of work can benefit AI ecosystems, but for this to happen, the knowledge of how to undertake anticipatory work needs to be available within the ecosystem.

One of the processes with the potential to draw knowledge into the AI ethics ecosystem is the engagement of all stakeholders. "Engagement" refers to activities that bring together different stakeholders in an open manner for a mutually informative exchange of ideas. The importance of engagement in science, research and technology development is long established (Arnstein 1969, Hart et al. 2009, Bickerstaff et al. 2010, Van Est 2011, Boulton et al. 2012). Certain aspects of engagement are also well established in technical disciplines, for example in the form of user engagement or user experience research, which form part of computer science, the parent discipline of AI (Haenlein and Kaplan 2019). However, in order to undertake engagement activities in a way that is ethically sensitive and can contribute to an AI ecosystem so as to promote human flourishing, they need to be employed carefully. Engagement in science and technology development is sometimes limited to exercises for the public understanding of science, which aim to inform the public about scientific insights or technical achievements. There is nothing wrong with such exercises, but they are only one part of public engagement, which, in order to live up to ethical expectations, needs to facilitate and open two-way communication, with researchers and other stakeholders being willing to engage, listen and respond to each other and take positions seriously. If this is not done in an inclusive manner, important knowledge to be gained about the AI ethics ecosystem might be lost.

While such an open engagement process promises both better understanding of the ecosystem through a broadening of the knowledge base and higher levels of acceptability of the resulting research and technologies, there is no guarantee that these will be achieved. Public debates about science, research and technology in many other areas, such as genetically modified organisms, nuclear energy and nanotechnology, show that engagement activities can be highly charged and conflictual (Van Est 2011). The potential for fundamental disagreements on underlying values, aims or desired outcomes pervades all such stakeholder engagements, whether high-profile at a national or international level or conducted at a local or organisational level.

All of these different aspects of knowledge need to exist in a practical and applicable form. It is not sufficient to have them in repositories that are not accessible or not used. Among the requirements for shaping AI ecosystems is thus that the knowledge base be accompanied by and to a large extent realised by a corresponding capacity to *apply* the knowledge. Capacity building is therefore a further key requirement: the different stakeholders need to not only recognise the legitimacy of different types of knowledge but be willing to engage with different knowledges and, ideally, develop their own capacities in applying these different knowledges. As Coeckelbergh (2020: 179) puts it, "if engineers learn to do things with texts and humanities people learn to do things with computers, there is more hope for a technology ethics and policy that works in practice". Of course, there are other stakeholders involved in AI ecosystems besides engineers and humanities specialists, and other knowledge domains besides "things with text" and "things with computers". But the general sentiment, that people need to be willing to gain new insights and learn new skills, is undoubtedly true.

The question of how this may be achieved brings us to the third group of requirements for shaping an AI ecosystem for human flourishing: the question of system governance.

7.2.3 Governance Principles of AI Ecosystems

The characteristics of innovation ecosystems and the resulting challenges for shaping AI ecosystems to promote human flourishing call for approaches to the governance of these systems that are sensitive to them. I am using the term "governance" here to denote all activities and processes that are put in place to facilitate and regulate the behaviour of the members of the AI ecosystem and the relationship of the ecosystem to its broader environment. The term frequently refers to structures and processes within organisations, whereas at a higher level the term "regulation" is used (Braithwaite and Drahos 2000). However, "governance" is increasingly used to describe a much broader array of

> processes of governing, whether undertaken by a government, market, or network, whether over a family, tribe, formal or informal organization, or territory, and whether through laws, norms, power, or language. (Bevir 2012: 1)

The term also refers to specific localised ways of organising (or governing) particular issues, as in data governance (Khatri and Brown 2010) or information governance (ISO 2008), rendering them suitable to describe ways of dealing with AI ecosystems that cover many societal actors and activities.

A key requirement for the governance of AI ecosystems is flexibility. We have seen that the members of an AI ecosystem are in complex and non-linear relationships. In addition, the technologies, organisations, social dynamics and other aspects of the ecosystem can change rapidly. Any governance structure therefore needs to be able to react flexibly to change. Kuhlmann et al. (2019) use the term "tentative governance" to describe this flexibility. They consider governance to be tentative

> when it is designed, practiced, exercised or evolves as a dynamic process to manage interdependencies and contingencies in a *non-finalizing* way; it is prudent (e.g. involving trial and error, or learning processes in general) and preliminary (e.g. temporally limited) rather than assertive and persistent. Tentative governance actors seek flexibility and act incrementally. (Kuhlmann et al. 2019: 1093, emphasis in original).

Such tentative governance needs to provide spaces for actors to learn and develop understanding of technologies, their use and their evaluation. It must be based on and foster communication between stakeholders. It should also allow for the acknowledgement of mistakes and have the ability to reverse or change course where initial assumptions prove to be wrong or where new insights or consensus emerge.

AI ecosystem governance should also be open to conflict and disagreement and be able to deal with those constructively. As Genus and Stirling (2018) rightly point out, responsible engagement with technology requires what they call "Collingridge qualities" (see Collingridge 1981), namely inclusion, openness, diversity, incrementalism, flexibility and reversibility. In many cases these can be better helped by exploring disagreement and dissensus than by engineering consensus.

The governance of AI ecosystems should be sensitive to the motivations and incentives of the members of the ecosystem. It needs to carefully balance the possible and expected benefits of AI with the possible and expected downsides. This requires an ability to draw on the knowledge and capacity described earlier, to evaluate developments and to put in place incentives and sanctions that reinforce developments that are desirable and promote human flourishing.

The AI ecosystem does not exist in a vacuum, and its governance should therefore be linked to existing governance structures. Questions regarding the extension of liability legislation to allow it to cover AI as currently discussed at the European level and elsewhere are one category of questions related to the extension of existing governance structures to include AI.

The consideration of existing governance structures is important to ensure the consistency of overlapping governance regimes, which, in turn, is important for the success of any governance efforts. Elsewhere (Stahl 2013, Stahl et al. 2019) I have introduced the concept of meta-responsibility as an important part of responsible research and innovation (RRI). This idea arises from the existence of networks of responsibility (Timmermans et al. 2017) which govern the practices of science, research and innovation. RRI as a meta-responsibility aims to shape, maintain, develop, coordinate and align existing and novel research- and innovation-related processes, actors and responsibilities with a view to ensuring desirable and acceptable research outcomes. This idea is relevant to AI ecosystems as well. AI ecosystems build on and incorporate many existing responsibilities and governance structures. In order for these ecosystems to be successful and to promote human flourishing, it is not necessary to re-invent principles of governance; they should rather be carefully developed, to help existing governance structures and responsibility relationships work effectively. In order to promote AI ecosystems that are conducive to human flourishing, we do not need to re-invent the wheel, but we need to make sure that the many wheels that already exist point in a roughly similar direction and that there is a strong and legitimate process that allows this direction to be determined.

Figure 7.2 presents a summary of the main points discussed in this section and aims to answer the question: which characteristics should an intervention into AI ecosystems have, to be likely to deal with ethical aspects successfully? These are necessary requirements, but may well not be the only ones, and are very unlikely to be sufficient on their own. They should nevertheless be useful in reviewing and evaluating practical interventions, policies and governance structures and could help

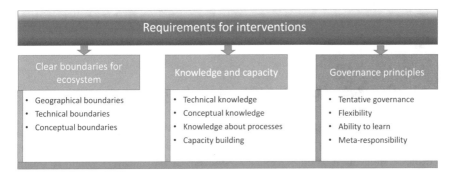

Fig. 7.2 Requirements for interventions into AI ecosystems

to improve those. They thus contribute to the types of recommendations which I outline in the next section.

7.3 Shaping AI Ecosystems

The proposals that follow are not ready-made recommendations that can be implemented as is to solve the ethical issues of AI. Apart from the fact that this is impossible because ethical issues do not normally lend themselves to simple resolution, I also lack the space in this book and the detailed knowledge of the various domains of expertise that would be required to develop detailed implementation plans.

What I am trying to do is to highlight some key aspects of governance, mitigation and interventions that have a high likelihood of making a positive contribution to the aim of shaping AI ecosystems in desired ways.

The audience envisaged for these proposals includes decision-makers who can shape aspects of the AI ecosystem, and also people who have an interest in and form part of the debate: policymakers as well as researchers and users of AI in public and private organisations, the media and civil society representatives.

AI ecosystems exist world-wide but are often geographically defined, subject to the laws of a particular jurisdiction, and can be sector-specific. My suggestions aim to be broadly applicable across geographies and jurisdictions, but they should be checked for local applicability. Moreover, most of the work that underpins my thinking was done in the UK and EU contexts, funded by European research grants. Many of the questions that influence how AI ecosystems can and will be shaped are being hotly debated at the time of writing (the European summer of 2020). At the European level in particular there is a set of proposals from the European Commission, the European Parliament and high-level expert groups. While this discussion is therefore clearly tilted towards Europe, I believe that the principles outlined are valid – or at least of interest – beyond Europe.

The sub-sections below focus on proposals for actions intended to ensure that AI ecosystems are conducive to human flourishing.

7.3.1   Conceptual Clarification: Move Beyond AI

Any successful intervention that promotes human flourishing in AI ecosystems needs to clarify which concept of AI is being used. While this is trivially obvious, it is also a difficult suggestion to implement. The concept of AI remains difficult to define and, as my discussion in Chapter 2 indicates, that there are fundamentally different technical artefacts and socio-technical systems. I am not suggesting that it would be possible to achieve conceptual divergence by decree. Rather, I believe that any intervention that aims to affect AI ecosystems must state clearly what it intends to cover. The multitude of meanings of AI and the lack of an agreed definition throw up serious doubts about the possibility of regulating AI (Stone et al. 2016).

Depending on the chosen concept of AI, it is possible that only a very specific part of the ecosystem will be affected, and the consequences for the overall ecosystem may be limited. For example, interventions that aim at a particular set of technologies, such as machine learning, will be limited to addressing effects that are clearly linked to these technologies, such as algorithmic biases or biases that arise from the development of models based on data containing biases. Choosing a narrow area of AI has the advantage of allowing for the definition of a closely circumscribed target for an intervention which then has a high likelihood of being successfully implemented. The disadvantage of such an approach is that interventions of this kind are not likely to have a major effect across broader AI ecosystems that are based on broader definitions of AI. The current approach by the European Commission (2020c) as outlined in its White Paper seems to pursue this strategy. While the definitions of AI in the document are broader, the aim seems to be to address the specifics of machine learning. My claim here is that the definition and the target of the intervention need to align.

The clear use of terminology is thus an important recommendation for anybody aiming to intervene in an AI ecosystem. However, competing understandings of the term "artificial intelligence" might lead to confusion. This is partly owing to the varying definitions, but also partly to the emotive connotations of AI. Humans tend to think of themselves as intelligent, and AI therefore has the potential to threaten our perceptions of ourselves. Popular culture and fiction have developed a range of easily recognisable tropes of AI, such as the evil robot, which colour the perception of actual and expected technologies. There can be little doubt that this is one of the causes of the high profile of the AI debate, but it is also its Achilles heel.

As a general suggestion I would therefore advocate using the term AI sparingly and moving beyond it to terms that better capture the particular technology or technologies in question. If an intervention is aimed at a specific aspect of AI, such as machine learning, then that should be made explicit. If the aim is to cover the broader ecosystem and address issues arising from life in a digital world, then a broader term would be more suitable.

My suggestion would be to speak of something like "emerging digital technologies", which probably covers the broad range of technologies of potential relevance, from narrow AI to neuromorphic computing, quantum computing, the internet of things, robotics and future digital technologies. This term will of course still have to be clearly defined, but the word "emerging" indicates that it is a moving target. It is also a much less emotive term than "artificial intelligence", probably not as threatening and likely to attract less attention. Considering the current hype around AI, I would suggest that a lower level of attention might be a good thing, as it might allow more careful deliberation in the planning of possible interventions.

7.3.2 Excellence and Flourishing: Recognise Their Interdependence

The previous suggestion pointed to the delimitation of the AI ecosystem in terms of the concepts and the technologies involved, whereas this one focuses on the conceptual clarity of the normative dimension. An intervention in an AI ecosystem requires the clarification of the purpose of the intervention: what is the intended outcome and why is the intervention deemed desirable?

In Chapter 3 I discussed the different purposes of AI that pervade much of the AI policy literature: economic benefits, social control and human flourishing. While this is an artificial and analytic distinction, I believe that a clear statement of purpose would be helpful for most interventions.

An intervention into an AI ecosystem for ethical purposes should explicitly state the ethical intention. I have advanced the concept of human flourishing as a broad term covering many different ethical positions and traditions. However, there are many other terms that could denote similar aims, such as fairness, equality, dignity and more specific aims such as the pursuit of the SDGs or the promotion of human rights. The point is that such a commitment is important to ensure that the intervention into the ecosystem can be monitored and assessed accordingly, and it should therefore be made explicit.

An important aspect of the commitment to an ethical position is its relationship to the technical environment. There is broad agreement that national and regional policymakers have a role in developing the technical AI ecosystem. That role tends to include measures such as the strengthening of educational pathways leading to an AI-proficient workforce, support for venture capital and other means of financing new AI-driven business models, and research and innovation structures that provide funding for AI-related research. The strengthening of the AI ecosystem is often also seen as requiring the creation of national or regional AI champions or centres of excellence that serve as hubs to bring together knowledge and expertise. In the UK this role is fulfilled by the Alan Turing Institute. The European Commission proposes the creation of networks of excellence at leading universities for a similar purpose.

From my perspective it is important to establish that the ethical dimension should be understood as an integral part of the technical AI ecosystem, not as an add-on. For AI to become truly geared towards human flourishing, it must overcome the current division between scientific excellence and ethical reflection and recognise that scientific excellence cannot be truly excellent if it does not consider the social and ethical consequences of the development, deployment and use of the technology.

At first sight, this might sound rather trivial, but it would require a far-reaching reconceptualisation of ethics in science, research and technology development. At present, ethics in research and innovation tends to be focused on research ethics, which typically takes the form of a review at the start of a project on the basis of biomedical research ethics principles. My suggestion is to fundamentally rethink the relationship of AI research and innovation and ethics. Ethics in the broad sense of promoting human flourishing should be an integral part of scientific excellence. This would mean that aspects such as intended consequences, detailed risk analysis and contingency plans that cover known or expected ethical issues would form part of the scientific evaluation of proposals and determine which ideas are seen as excellent and thus worthy of being funded.

7.3.3   Measurements of Flourishing: Understanding Expected Impacts

An important ingredient for ensuring that AI ecosystems drive towards human flourishing is the ability to reflect on the successes of prior interventions and use them as a basis for steering the ecosystem. This requires an understanding of the intended and real impacts of the activities in the ecosystem. With regard to the desire to promote human flourishing, it calls for ways of understanding human flourishing in practice.

At present there is no agreed-upon methodology or set of criteria that can be used to assess how the consequences of the development and use of AI have affected or are affecting human flourishing. Or, to put it differently, we need to have measures for human flourishing and these need to be applicable to AI ecosystems. However, we do not need to start from scratch in developing such measurements.

There are well-accepted mechanisms that provide at least parts of what is required. One established method for measuring human flourishing is the Human Development Index (UNDP n.d.). This set of measures was inspired by the Capability Approach (Sen 1993, Alkire 2002, Nussbaum 2011), which set out to move beyond the measurement of gross domestic product and instead evaluate the quality of human development. It has been adopted by the United Nations Development Programme as a key measure and has also been successfully applied to ICT (Johnstone 2007, Zheng and Stahl 2011). Hence, it seems plausible that it could easily be tailored to cover the specific aspects of AI.

A similar set of ideas has driven the development of the SDGs, which currently seem more prominent with regard to AI, notably in the "AI for Good" discourse (Taddeo and Floridi 2018). Orienting AI development, deployment and use towards the achievement of the SDGs is a key component of AI for Good as promoted by the International Telecommunications Union (ITU n.d.). The idea of moving towards the SDGs is currently not very contentious, even though in practice it may not be trivial and could lead to trade-offs between different ethically relevant goals (Ryan et al. 2020)

A further approach that also has the advantage of being based on recognised international agreements is the focus on human rights. As discussed in Chapter 4, there are already a number of proposals on finding ways of applying SDGs or human rights to AI (Raso et al. 2018, Latonero 2018, Commissioner for Human Rights 2019).

All of these approaches appear to have a high potential for being applicable to AI and providing ways to structure discussion and understanding of the impact of interventions in the AI ecosystem. A clearer understanding of their respective strengths and weaknesses would be helpful in deciding which ones might be most appropriate in which AI ecosystems.

The question of human flourishing and the influence that AI can have on this is not easy to resolve. The pointers in this section to the Human Development Index, the SDGs and human rights measures are meant to provide indications of how such influence may be achieved. Trying to measure something as complex and multi-faceted as flourishing raises many challenges. Some of these are methodological and epistemological, revolving around the questions: what can we measure and how can it be measured? The very term "measure" suggests a quantitative approach, and the degree to a complex qualitative term such as flourishing can be captured using quantitative measures is open to debate. The challenges go even further and touch on the heart of ethics, on the question: is it suitable at all to even try to measure human flourishing?

This book cannot offer a comprehensive answer to that question. However, it can point to the fact that we live in a world where measurements drive many policies and behaviours. An ability to express whether a particular aim, such as human flourishing, has been achieved, or whether an activity or process can lead to progress in the direction of this aim, would therefore help engage decision-makers who are used to this type of discourse. Developing measurements is highly ambitious, and it is very unlikely that we will ever be able to measure human flourishing comprehensively. But the benefits of having some, albeit imperfect, measures may well be worth the disagreements that these measures are likely to evoke.

7.3.4 AI Benefits, Risks and Capabilities: Communication, Knowledge and Capacity Building

At the heart of any attempt to shape AI ecosystems and move them in the direction of human flourishing must be an understanding of the benefits and risks of the technologies and the capabilities they can bestow on users. The fast-moving nature of AI means that this knowledge may lose its currency quickly, which is why I suggest that an AI knowledge base is a requirement for the successful shaping of AI ecosystems.

Such a knowledge base would no doubt be based on existing structures of knowledge and learning, including academic publications and databases, and web resources. A key role in establishing and maintaining this knowledge base would be filled by the centres of excellence – those that are already established and the new centres or network structures that are being developed. In addition, several international organisations, such as the OECD, UNESCO and the European Commission, are developing databases, observatories etc. to capture the discourse. Standardisation bodies have an important role to play in collecting available knowledge and facilitating consensus on key aspects.

One key suggestion I would like to make in this respect mirrors the one in the section on excellence and flourishing (Section 7.3.2), namely, to ensure that no artificial divide is imposed between scientific knowledge and ethical and social understanding. This means that AI centres of excellence should include excellence in the ethics of AI, a position that the Alan Turing Institute in the UK, for example, has already adopted. Similarly, while there is no doubt scope for specialised standardisation on ethics and AI, as the IEEE P7000 family of standards shows, it would be desirable for technical AI standards to refer to and include ethical aspects.

The AI knowledge base needs to be scientifically sound and reliable, but it must also be visible, communicated and understood, which implies the need for educational activities, from primary education all the way up to post-doctoral work. This, in turn, calls for reviews of national and disciplinary curricula, the development of learning support and the creation of teaching capacity.

The further dissemination and practical societal usefulness of this knowledge will depend on whether it can be conveyed in a simple and comprehensible manner. One approach to this is to develop labels and certificates for AI systems, comparable to well-established labels such as those codifying energy consumption, the nutritional content of food and environmental sustainability. It may help to use existing categorisations of AI, such as the six levels of autonomy – levels 0 to 5 (SAE 2018) – that are used for autonomous vehicles to convey relevant aspects of benefit and risk. Such relatively simple categorisations of important aspects of AI may help visually represent possible benefits and risks and thus support balanced decision making.

7.3.5 Stakeholder Engagement: Understanding Societal Preferences

The suggestions above assume that there is a position on AI that allows us to determine which uses and applications of technology are desirable and acceptable, that there is some sort of agreement on what counts as flourishing or which benefits warrant taking particular risks. While I believe that one can indeed find much consensus on many of these questions, at least within specific communities and states, there will always be new or borderline phenomena that are less clearly understood and give rise to different interpretations and evaluations.

The complexity of AI and other emerging digital technologies, both in terms of their technical capacities and in relation to societal outcomes and impact, means that it is unlikely that these questions will be easy to settle. Furthermore, in many cases they will fall into line with existing societal disagreements, e.g. with regard to what counts as just distribution or what a state can reasonably require its citizens to do.

A full understanding of what counts as an ethical issue related to AI, why it counts and what, if anything, could or should be done about it therefore calls for societal debates that allow stakeholders to come together and debate these questions. As a consequence, the ethics of AI cannot be a topic that is dealt with by technical and ethical experts alone: it calls for broader stakeholder engagement.

To a large extent the political processes that exist in a democratic state can take care of this task and provide means for the expression of divergent opinions and legitimate decisions concerning desirable actions. In order for AI ecosystems to be steered towards human flourishing, they will therefore need mechanisms that institutionalise stakeholder engagement activities that give stakeholders a voice and allow them to contribute meaningfully to collective decision-making. Appropriate recommendations and policies seem to call for a multi-stakeholder approach that brings together relevant stakeholders in an inclusive manner to move towards human flourishing or, as Cath et al. (2016: 18) put it, to deliver a "good AI society".

This is much easier said than done. There are many potential pitfalls in stakeholder engagement. Such activities need to be carefully defined, planned and executed to avoid their being hijacked by particular interests (Wehling 2012). They need to be aligned with existing democratic processes. There are difficult questions about the frequency and intensity of stakeholder engagements which have to do with the costs they incur and whether they can truly claim to represent stakeholder opinions (Wynne 2006, Felt and Fochler 2010). Notwithstanding these potential problems and downsides, however, it is difficult to see how AI ecosystems can properly understand the ethical issues they face and acceptable ways of dealing with them unless they have appropriate ways of consulting stakeholders.

7.3.6 Responsibility for Regulation and Enforcement: Defining the Central Node(s) of the AI Ecosystems

My last suggestion for shaping the AI ecosystem has a bearing on all the other suggestions. It relates to to the question of where responsibility lies for planning, realising, implementing and enforcing the suggestions made here and elsewhere. This refers to the concept of meta-responsibility, i.e. the question of who or what is responsible for ensuring that individuals, organisations and states understand their responsibilities and fulfil them.

I believe that a key condition for any suggestion or recommendation to be successful is the ability to answer the question: who or what is responsible for implementing it? At the national or international level this relates to the question of whether there should be a regulator for AI and what form it should take. At the European level we observe several opinions on this. The European Commission (2020a) is in favour of a strengthening of the network of existing regulators, whereas the European Parliament (2020b) has proposed the creation of a European Agency for Artificial Intelligence.

I will not comment in detail on this discussion but would like to point to some aspects that should be considered when seeking a way forward. Some structure or body has to take responsibility for many aspects of ethics in AI ecosystems. There must be a place where conceptual positions are collected and defined. The knowledge base and ways of measuring and assessing technologies and their impact need an institutional home, which such a network of existing regulators or a new regulator could offer.

In the UK the non-profit organisation Doteveryone has published a report (Miller and Ohrvik-Stott 2018) on regulating responsible technology which contains a strong analysis of the challenges and proposes the creation of a central hub to guide and support a number of activities. This report employs the ecosystem metaphor of digital technologies and builds on it to explore ways in which entire ecosystems can be governed to serve society. At the core of this report's recommendations is the creation of what the authors call the Office of Responsible Technology.

The proposed office is explicitly not a regulator for AI. Such a regulator would fall victim to the problem of the lack of clarity in defining AI and might end up as a regulator for everything. Instead, it would be set up as an organisation to support and strengthen existing regulators, such as data protection authorities and financial or other sectoral regulators. These existing regulators are mostly well established and best placed to deal with particular applications, but they often lack knowledge and experience specific to AI or other emerging technologies. The Office for Responsible Technology is therefore described as an organisation that works with regulators and provides the technology-specific knowledge and expertise that they lack.

The Doteveryone report envisages another set of tasks for this office that aligns directly with some of the suggestions I made earlier. It is designated as the home of public engagement, both for the exchange of information and as the place where a vision for technology and society is developed. The report also sees the Office for

Responsible Technology as the body responsible for ensuring that redress procedures exist and are usable.

Some aspects of these proposals are debatable. I do not think that all the various tasks proposed for the Office for Responsible Technology need to be located in one organisation. Such a concentration of tasks might make it a large, cumbersome and bureaucratic institution. At the same time it is clear that this idea has traction, as can be seen from the current European discussion of a potential regulator as well as from the fact that organisations are starting to emerge that cover at least parts of this remit, such as the UK's Centre for Data Ethics and Innovation.[1]

Such an organisation is certainly needed at the political level, whether it be called a regulator, an office, a centre or something else. It should not, however, set out to regulate all of AI, if for no other reason than that it is difficult to define the term. Rather, it should have a remit that covers emerging (digital) technologies and should support existing regulatory structures and processes. This would be a subject of meta-responsibility, i.e. the office would be the organisation responsible for ensuring that technology-related responsibilities are clearly defined and can be fulfilled.

It is worth pointing out that this principle of having a subject of meta-responsibility is not confined to the political level and to national or regional AI ecosystems. A similar organisation or role will be required in other ecosystems, to ensure that there is a mechanism for all ecosystem members to access knowledge, develop capacities, receive guidance and provide input into governance structures. At the level of an organisation this could be a trusted position with responsibility for AI in that organisation. The incumbent could be called the AI officer or, perhaps better, the digital officer. This could be developed in a similar way to the data protection officer, a role that is mandated for European organisations to ensure that data protection requirements are met. The data protection officer works for and is paid by the organisation but has a responsibility defined with regard to data protection requirements, not organisational needs. In the case of a conflict between these, the perspective of the data protection officer is broader than that of the organisation. A similar role with regard to AI could be of crucial importance for the governance of organisational AI ecosystems, which could be a cornerstone of larger and overarching ecosystems. Where appropriate, such roles could also be combined, so that in a company that makes significant use of AI but is not centrally geared towards AI, the data protection officer could simultaneously serve as digital officer.

The suggestions put forward in this section offer examples of the types of actions and interventions that I believe can help move AI ecosystems in a direction conducive to human flourishing, although implementing them will require more thought and detail. The exact form such actions and interventions eventually take will be the subject of further discussion, but my suggestions go some way towards addressing the challenges of AI ecosystems and are consistent with the requirements for interventions that I set out earlier. They could determine the shape of a future-oriented governance framework. Such a framework needs flexibility to ensure that future technologies are accommodated and must open up productive discussion between

---

[1] https://www.gov.uk/government/organisations/centre-for-data-ethics-and-innovation

stakeholders in organisations and countries and internationally to determine how the various AI and related ecosystems are to be developed.

References

Alkire S (2002) Valuing freedom: Sen's capability approach and poverty reduction. Oxford University Press, Oxford

Anand P, Santos C, Smith R (2008) Poverty, capabilities and measurement, Oxford University Press, Oxford. https://www.poverty.ac.uk/system/files/poverty_capabilities-and-measurement.pdf. Accessed 29 Oct 2020

Arnstein SR (1969) A ladder of citizen participation. J Am Inst Plann 35:216–224. https://doi.org/10.1080/01944366908977225

Bevir M (2012) Governance: a very short introduction. Oxford University Press, Oxford

Bickerstaff K, Lorenzoni I, Jones M, Pidgeon N (2010) Locating scientific citizenship: the institutional contexts and cultures of public engagement. Sci Technol Hum Values 35:474–500. https://doi.org/10.1177/2F0162243909345835

Boulton G, Campbell P, Collins B et al (2012) Science as an open enterprise. Royal Society, London. https://royalsociety.org/-/media/policy/projects/sape/2012-06-20-saoe.pdf. Accessed 29 Oct 2020

Braithwaite J, Drahos P (2000) Global business regulation. Cambridge University Press, Cambridge

Brey P (2011) Anticipatory technology ethics for emerging IT. In: Mauger J (ed) CEPE 2011: crossing boundaries. INSEIT, Nice, France, pp 13–26

Buckingham W (n.d.) Vulnerability and flourishing—Martha Nussbaum. https://gohighbrow.com/vulnerability-and-flourishing-martha-nussbaum/. Accessed 3 Nov 2020

Cath CJN, Wachter S, Mittelstadt B, Taddeo M, Floridi L (2016) Artificial intelligence and the 'good society': the US, EU, and UK approach. Social Science Research Network, Rochester, NY

Coeckelbergh M (2020) AI ethics. The MIT Press, Cambridge, MA

Collingridge D (1981) The social control of technology. Palgrave Macmillan, London

Commissioner for Human Rights (2019) Unboxing artificial intelligence: 10 steps to protect human rights. https://rm.coe.int/unboxing-artificial-intelligence-10-steps-to-protect-human-rights-reco/1680946e64. Accessed 6 Oct 2020

Cuhls K (2003) From forecasting to foresight processes: new participative foresight activities in Germany. J Forecast 22:93–111. https://doi.org/10.1002/for.848

EDPS (2020) EDPS opinion on the European Commission's White Paper on Artificial Intelligence—a European approach to excellence and trust (Opinion 4/2020). European Data Protection Supervisor, Brussels

European Commission (2020a) Report on the safety and liability implications of artificial intelligence, the internet of things and robotics. European Commission, Brussels. https://ec.europa.eu/info/files/commission-report-safety-and-liability-implications-ai-internet-things-and-robotics_en. Accessed 22 Sept 2020

European Commission (2020b) Shaping Europe's digital future. Communication from the Commission to the European Parliament, the Council, the European Economic and Social Committee and the Committee of the Regions. European Commission, Brussels. https://ec.europa.eu/info/sites/info/files/communication-shaping-europes-digital-future-feb2020_en_3.pdf. Accessed 6 Oct 2020 [MAKE THIS 2020c]

European Commission (2020c) White paper on artificial intelligence: a European approach to excellence and trust. European Commission, Brussels. https://ec.europa.eu/info/sites/info/files/commission-white-paper-artificial-intelligence-feb2020_en.pdf. Accessed 22 Sept 2020

European Parliament (2020a) Draft report with recommendations to the Commission on a civil liability regime for artificial intelligence. European Parliament, Committee on Legal

Affairs. https://www.europarl.europa.eu/doceo/document/JURI-PR-650556_EN.pdf. Accessed 6 Oct 2020

European Parliament (2020b) Draft report with recommendations to the Commission on a framework of ethical aspects of artificial intelligence, robotics and related technologies. European Parliament, Committee on Legal Affairs. https://www.europarl.europa.eu/doceo/document/JURI-PR-650508_EN.pdf. Accessed 25 Sept 2020

Expert Group on Liability and New Technologies (2019) Liability for artificial intelligence and other emerging digital technologies. Publications Office of the European Union, Luxembourg. https://op.europa.eu/en/publication-detail/-/publication/1c5e30be-1197-11ea-8c1f-01aa75ed71a1/language-en/format-PDF. Accessed 23 Sept 2020

Felt U, Fochler M (2010) Machineries for making publics: inscribing and de-scribing publics in public engagement. Minerva 48:219–238. https://doi.org/10.1007/s11024-010-9155-x

Flick C, Zamani ED, Stahl BC, Brem A (2020) The future of ICT for health and ageing: unveiling ethical and social issues through horizon scanning foresight. Technol Forecast Soc Change 155:119995. https://doi.org/10.1016/j.techfore.2020.119995

Floridi L, Strait A (2020) Ethical foresight analysis: what it is and why it is needed? Minds Mach. https://doi.org/10.1007/s11023-020-09521-y

Gal D (2020) China's approach to AI ethics. In: Elliott H (ed) The AI powered state: China's approach to public sector innovation. Nesta, London, pp 53–62

Genus A, Stirling A (2018) Collingridge and the dilemma of control: towards responsible and accountable innovation. Res Policy 47:61–69. https://doi.org/10.1016/j.respol.2017.09.012

Haenlein M, Kaplan A (2019) A brief history of artificial intelligence: on the past, present, and future of artificial intelligence. Calif Manage Rev 61:5–14. https://doi.org/10.1177/2F0008125619864925

Hart A, Northmore S, Gerhardt C (2009) Auditing, benchmarking and evaluating public engagement. National Co-ordinating Centre for Public Engagement, Bristol

Harwell D (2020) Managers turn to surveillance software, always-on webcams to ensure employees are (really) working from home. The Washington Post. https://www.washingtonpost.com/technology/2020/04/30/work-from-home-surveillance/. Accessed 30 Oct 2020

Himanen P (2001) The hacker ethic, and the spirit of the information age. Random House, New York

Holzinger A, Biemann C, Pattichis CS, Kell DB (2017) What do we need to build explainable AI systems for the medical domain? arXiv:1712.09923 [cs, stat]

ISO (2008) BS ISO/IEC 38500:2008: Corporate governance of information technology. British Standards Institute, London

ITU (n.d.) AI for good global summit. International Telecommunications Union and XPRIZE Foundation. https://aiforgood.itu.int/. Accessed 30 Oct 2020

Johnstone J (2007) Technology as empowerment: a capability approach to computer ethics. Ethics Inf Technol 9:73–87. https://doi.org/10.1007/s10676-006-9127-x

Khatri V, Brown CV (2010) Designing data governance. Commun ACM 53:148–152. https://doi.org/10.1145/1629175.1629210

Klar R, Lanzerath D (2020) The ethics of COVID-19 tracking apps—challenges and voluntariness. Res Ethics 16:1–9. https://doi.org/10.1177/2F1747016120943622

Knight W (2020) AI can do great things—if it doesn't burn the planet. Wired. https://www.wired.com/story/ai-great-things-burn-planet/. Accessed 1 Nov 2020

Kriechbaum M, López Prol J, Posch A (2018) Looking back at the future: dynamics of collective expectations about photovoltaic technology in Germany & Spain. Technol Forecast Soc Change 129:76–87. https://doi.org/10.1016/j.techfore.2017.12.003

Kuhlmann S, Stegmaier P, Konrad K (2019) The tentative governance of emerging science and technology: a conceptual introduction. Res Policy 48:1091–1097. https://doi.org/10.1016/j.respol.2019.01.006

Latonero M (2018) Governing artificial intelligence: upholding human rights & dignity. Data Soc. https://datasociety.net/wp-content/uploads/2018/10/DataSociety_Governing_Artificial_Intelligence_Upholding_Human_Rights.pdf. Accessed 26 Sept 2020

Miller C, Ohrvik-Stott J (2018) Regulating for responsible technology—capacity, evidence and redress: a new system for a fairer future. Doteveryone, London

Nussbaum M (2000) Women and human development: the capabilities approach. Cambridge University Press, Cambridge

Nussbaum MC (2011) Creating capabilities: the human development approach. Harvard University Press, Cambridge MA

Raso FA, Hilligoss H, Krishnamurthy V et al. (2018) Artificial intelligence & human rights: opportunities & risks. Berkman Klein Center Research Publication No. 2018-6. http://dx.doi.org/10.2139/ssrn.3259344

Ryan M, Antoniou J, Brooks L et al (2020) The ethical balance of using smart information systems for promoting the United Nations' Sustainable Development Goals. Sustainability 12:4826. https://doi.org/10.3390/su12124826

SAE (2018) Taxonomy and definitions for terms related to driving automation systems for on-road motor vehicles J3016_201806. SAE International, Warrendale PA. https://www.sae.org/standards/content/j3016_201806/. Accessed 30 Oct 2020

Sardar Z (2010) The namesake: futures; futures studies; futurology; futuristic; foresight—what's in a name? Futures 42:177–184. https://doi.org/10.1016/j.futures.2009.11.001

Scott D, Rutty M, Peister C (2018) Climate variability and water use on golf courses: optimization opportunities for a warmer future. J Sustain Tour 26:1453–1467. https://doi.org/10.1080/09669582.2018.1459629

Sen A (1993) Capability and well-being. In: Nussbaum M, Sen A (eds) The quality of life. Clarendon Press, Oxford, pp 30–53

Stahl BC (2013) Responsible research and innovation: the role of privacy in an emerging framework. Sci Public Policy 40:708–716. https://doi.org/10.1093/scipol/sct067

Stahl BC, Timmermans J, Rainey S, Shaw M (2019) Ethics in innovation management as meta-responsibility: the practice of responsible research and innovation in human brain simulation. In: Chen J, Brem A, Viardot E, Wong PK (eds) The Routledge companion to innovation management, 1st edn. Routledge, New York, pp 435–454

Stahl BC, Wright D (2018) Ethics and privacy in AI and big data: implementing responsible research and innovation. IEEE Secur Priv 16:26–33. https://doi.org/10.1109/MSP.2018.2701164

Stilgoe J, Owen R, Macnaghten P (2013) Developing a framework for responsible innovation. Res Policy 42:1568–1580. https://doi.org/10.1016/j.respol.2013.05.008

Stone P, Brooks R, Brynjolfsson E et al (2016) Artificial intelligence and life in 2030: one hundred year study on artificial intelligence. Report of the 2015–2016 study panel. Stanford University, Stanford CA. http://ai100.stanford.edu/2016-report. Accessed 6 Sept 2016

Taddeo M, Floridi L (2018) How AI can be a force for good. Science 361:751–752. https://doi.org/10.1126/science.aat5991

Tenner E (1997) Why things bite back: predicting the problems of progress, new edn. Fourth Estate, London

Timmermans J, Yaghmaei E, Stahl BC, Brem A (2017) Research and innovation processes revisited: networked responsibility in industry. Sustainability 8:307–334. https://doi.org/10.1108/SAMPJ-04-2015-0023

UNDP (n.d.) Human development index (HDI). United Nations Development Programme. http://hdr.undp.org/en/content/human-development-index-hdi. Accessed 2 Nov 2020

Van Est R (2011) The broad challenge of public engagement in science. Sci Eng Ethics 17:639–648. https://doi.org/10.1007/s11948-011-9296-9

Wehling P (2012) From invited to uninvited participation (and back?): rethinking civil society engagement in technology assessment and development. Poiesis Prax 1–18. https://doi.org/10.1007/s10202-012-0125-2

Wynne B (2006) Public engagement as a means of restoring public trust in science—Hitting the notes, but missing the music? Commun Genet 9:211–220. https://doi.org/10.1159/000092659

Xu S, Zhu K, Gibbs J (2004) Global technology, local adoption: a cross-country investigation of internet adoption by companies in the United States and China. Electron Mark 14:13–24. https://doi.org/10.1080/1019678042000175261

Zheng Y, Stahl BC (2011) Technology, capabilities and critical perspectives: what can critical theory contribute to Sen's capability approach? Ethics Inf Technol 13:69–80. https://doi.org/10.1007/s10676-011-9264-8

Chapter 8
Conclusion

Abstract The conclusion briefly summarises the main arguments of the book. It focuses on the requirements for mitigation options to be used to address the ethical and human rights concerns of artificial intelligence. It also provides a high-level overview of the main recommendations brought forth in the book. It thereby shows how conceptual and empirical insights into the nature of AI, the ethical issues thus raised and the mitigation strategies currently being discussed can be used to develop practically relevant conclusions. These conclusions and recommendations help to ensure that AI ecosystems are developed and shaped in ways that are conducive to human flourishing.

Keywords Requirements for AI ethics · Recommendations for AI · AI governance · Ethics and AI ecosystems · AI regulation

Technology is part of human life. Its development and use have the potential to raise ethical concerns and issues – and this will not change. Ethics, understood as our struggle to determine what is right and wrong and our reflection on how and why we make such a distinction, is not subject to resolution. While we may agree on what is right and wrong in many cases, this agreement is always partial, temporary and subject to revision. We may, however, be able to agree on some general and abstract principles. In this book I have suggested that human flourishing is such a principle. If we agree on that, then we can think through what the application of the principle to a technology such as AI can mean. This exercise can help us understand the specific issues that arise, why they arise and how we can evaluate them. It can also help us think through what we can do about them, and may even help us resolve some of them to universal satisfaction.

Several aspects that I have focused on in this book can, I hope, make a novel and interesting contribution to the AI and ethics debate. I started by looking at the concept of AI. "Artificial intelligence" is not an innocent and morally neutral term. It is emotive because it points to a characteristic of humans (and to some degree of other animals) while implying that this characteristic can be artificially replicated. This implication has consequences for how we as humans see ourselves and our role

© The Author(s) 2021
B. C. Stahl, *Artificial Intelligence for a Better Future*,
SpringerBriefs in Research and Innovation Governance,
https://doi.org/10.1007/978-3-030-69978-9_8

in the world. Artificial intelligence is also often contrasted with human intelligence, implicitly suggesting or explicitly asserting that machines can or even should replace humans. Again, this touches deeply rooted views of what humans are.

In order to render the discussion more accessible, I have proposed a new categorisation of the AI debate. My suggestion is that we distinguish between three perspectives on AI: machine learning or narrow AI, general AI and converging socio-technical systems. These three perspectives on the technology are enlightening because they align with the categorisation of ethical issues on AI: first, ethical issues related to machine learning; second, general issues related to living in a digital world; and third, metaphysical issues posed by AI. These distinctions thus provide a better understanding and overview of AI ethics in a very busy and often overwhelming public and academic debate.

While these categorisations clarify the debate, they say very little about what could or should be done about the issues. One of the problems in this type of normative discussion is that it is unclear how recommendations or prescriptions can be justified. On what grounds could we say that technical applications should be developed, promoted, avoided or prohibited? Drawing on the idea of human flourishing allows a normative point of reference to be established that is consistent and compatible with the main ethical theories and can provide a framework for thinking about normative questions without presupposing substantive moral positions.

The idea of human flourishing has the added advantage of not requiring a strict distinction between ethics and law, both of which are normative constructs that could promote or inhibit flourishing. This is particularly important in light of the numerous existing legal and regulatory rules that already guide the development and use of technology, including AI.

Drawing on rich empirical work, I analysed ethical concerns and suggested interventions, mitigations and governance approaches to promote the benefits of AI and avoid or address its downsides.

One problem in the AI ethics discussion is its high level of complexity. Any attempt to match individual issues with stakeholders and mitigation options runs into several problems. First, the number of possible combinations of stakeholders, mitigations and ethical issues to be addressed is such that it is impractical to try to understand the field using such a straightforward approach. Second, and more important, the different components of the interaction are not independent, and an intervention in one part is likely to have consequences in another part. As this type of dynamic relationship lends itself to being described using a systems perspective, I have adopted the now widely used ecosystem metaphor and applied it to the AI discourse.

The question of what needs to be done to ensure that AI ecosystems are conducive to human flourishing was then tackled through the ecosystem metaphor. This led me to investigate, from an ethical perspective, the implications of using the ecosystem metaphor, a question that is not yet widely pursued in the AI field. In addition, I analysed the challenges that the ecosystem approach to AI and ethics raises and the requirements that any intervention would need to fulfil, and I concluded with suggestions to take the debate further and provide input into discussions.

The analysis pointed to three groups of requirements that interventions into AI ecosystems need to fulfil, in order to increase their chances of successfully promoting human flourishing:

- **Interventions need to clearly delineate the boundaries of the ecosystem**: Systems boundaries are not necessarily clear and obvious. In order to support AI ecosystems, the boundaries of the ecosystem in question need to be clearly located. This refers not only to geographical and jurisdictional boundaries, but also to conceptual ones, i.e. the question of which concept of AI is the target of intervention and which ethical and normative concepts are at the centre of attention.
- **Interventions need to develop, support, maintain and disseminate knowledge**: The members of AI ecosystems require knowledge, if they are to work together to identify ethically desirable future states and find ways of working towards those. AI as a set of advanced technologies requires extensive subject expertise in the technologies, their capacities and uses. In addition, AI ecosystems for human flourishing require knowledge about concepts and processes that support and underpin ethical reflections. And, finally, AI ecosystems need mechanisms that allow for these various bodies of knowledge to be updated and made available to members of those ecosystems who need them in a particular situation.
- **Interventions need to be adaptive, flexible and able to learn:** The fast-moving nature of AI-related innovation and technology development, but also of social structures and preferences as well as adjacent innovation ecosystems, means that any intervention into the AI ecosystem needs to incorporate the possibility and, indeed, likelihood of change. Governance structures therefore need to be flexible and adaptable. They need to be open to learning and revisions. They need to be cognisant of existing responsibilities and must build and shape these to develop the ecosystem in the direction of human flourishing.

These requirements are deduced from the nature and characteristics of AI innovation ecosystems. They are likely to have different weights in different circumstances and may need to be supplemented by additional requirements. They constitute the basis of the recommendations developed in this book. Before I return to these recommendations it is worth reflecting on future work.

The work described in this book calls for development in several directions. An immediate starting point is a better empirical understanding of the impact of AI and digital technologies across several fields and application areas. We need detailed understanding of the use of technologies in various domains and the consequences arising. We also need a much broader geographical coverage to ensure that the specifics of different nations, regions and cultures are properly understood.

Such empirical social science research should be integrated into the scientific and technical research and development activities in the AI field. We need a strong knowledge base to help stakeholders understand how particular technologies are used in different areas, which can help technical researchers and developers as well as users, deployers, policymakers and regulators.

The insights developed this way will need to be curated and made available to stakeholders in a suitable way. To a large extent this can be done through existing structures, notably the scientific publication process. However, issues of legislation, regulation and compliance require special gatekeepers who can lay claim not only to a high level of scientific and technical expertise, but also to normative legitimacy. The idea is not to install a tyranny of the regulator, but to establish ways that help stakeholders navigate the complexity of the debate and spaces in which organisations and societies can conduct a fruitful debate about desirable futures and the role that technologies should play in them.

The discussion of the ethics of AI remains high-profile. Numerous policy and regulatory proposals are likely to be implemented soon. The causes of the high level of attention that AI receives remain pertinent. The technologies that constitute AI continue to develop rapidly and are expected to have a significant social and economic impact. They promise immense benefits and simultaneously raise deep concerns. Striking an appropriate balance between benefits and risks calls for difficult decisions drawing on expertise in technical, legal, ethical, social, economic and other fields.

In this book I have made suggestions on how to think about these questions and how to navigate the complexity of the debate, and I have provided some suggestions on what should be done to facilitate this discussion. These recommendations have the purpose of moving AI ecosystems in the direction of human flourishing. They satisfy the three requirements listed above, namely to delineate the ecosystems boundaries, to establish and maintain the required knowledge base and to provide flexible and adaptive governance structures. In slightly more detail (see Chapter 7 for the full account), the recommendations are:

- **Conceptual clarification: Move beyond AI (7.3.1)**
  The concept of AI is complex and multi-faceted (see Chapter 2). The extent of the ecosystems concerned and the ethical and human rights issues that are relevant in them depend to a large degree on the meaning of the term "artificial intelligence". Any practical intervention should therefore be clear on the meaning of the concept. It will often be appropriate to use a more specific term, such as "machine learning" or "neural network", where the issues are related to the characteristics of the technology. It may also be appropriate to use a wider term such as "emerging digital technologies", where broad societal implications are of interest.
- **Excellence and flourishing: Recognise their interdependence (7.3.2)**
  In the current discussion of AI, including some of the policy-oriented discourses, there is a tendency to distinguish between the technical side of AI, in which scientific and technical expertise is a priority, and the ethical and human rights side. This blurs the boundaries of what is or should be of relevance in an AI ecosystem. The recommendation points to the fact that scientific and technical excellence must explicitly include social and ethical aspects. Work on AI systems that ignores social and ethical consequences cannot be considered excellent.

- **Measurements of flourishing: Understanding expected impacts (7.3.3)**
  In order to react appropriately to the development, deployment and use of AI, we must be able to understand the impact they can be expected to have. It is therefore important to build a knowledge base that allows us to measure (not necessarily using quantitative metrics) the impact across the range of AI technologies and application areas. While it is unlikely to be possible to comprehensively measure all possible ethical, social and human rights impacts, there are families of measurements of aspects of human flourishing that can be applied to AI, and these need to be developed and promoted.
- **AI benefits, risks and capabilities: Communication, knowledge and capacity building (7.3.4)**
  The knowledge base of AI ecosystems needs to cover the technical side of AI technologies, to ensure that the risks and potential benefits of these technologies can be clearly understood. This knowledge, combined with the measures of human flourishing in the preceding recommendation, is required for a measured view of the impact of AI systems and a measured evaluation of their benefits and downsides. This knowledge base that AI ecosystems must be able to draw on, in order to make justifiable decisions on AI, is dynamic and can be expected to evolve quickly. It therefore needs to develop mechanisms for the regular updating and development of expertise and means of disseminating it to those who need it.
- **Stakeholder engagement: Understanding societal preferences (7.3.5)**
  The broad and all-encompassing nature of AI and its possible impacts means that decisions shaping the development, deployment and use of AI and hence its societal impact must be subject to public debate. Established mechanisms of representative democracy have an important role to play in guiding AI governance. However, the dynamic and complex nature of the field means that additional mechanisms for understanding the views and perceptions of stakeholders should be employed. Involving stakeholders in meaningful two-way communication with researchers, scientists and industry has the advantage of increasing the knowledge base that technical experts can draw on, as well as improving the legitimacy of decisions and policies resulting from such stakeholder engagements.
- **Responsibility for regulation and enforcement: Defining the central node(s) of AI ecosystems (7.3.6)**
  AI ecosystems do not develop in a vacuum but emerge from existing technical, social, legal and political ecosystems. These ecosystems have developed a plethora of mechanisms to attribute responsibility with a view to ensuring that the risks and benefits of emerging technologies are ascribed appropriately. The emergence of AI ecosystems within these existing environments means that existing roles and responsibilities need to be suitably modified and developed. This calls for a way of coordinating the transition to AI ecosystems and integrating them into established contexts. The shifting networks of responsibilities that govern emerging technologies will therefore need to evolve ways of developing formal and informal governance structures and monitoring their implementation. This calls for the establishment of central nodes (e.g. regulators, agencies, centres of excellence)

that link, guide and oversee AI ecosystems, and relevant knowledge and structures to ensure the technologies contribute to human flourishing.

I hope that this book and the recommendations that arise from it help strengthen the debate on AI and ethics. The book aims to support the appropriate shaping of AI ecosystems. In addition, its message should reach beyond the current focus on AI and help to develop our thinking on the technologies that will succeed AI at the centre of public attention.

Humans are and will remain tool-using animals. The importance of technical tools will increase, if anything, in times of ubiquitous, pervasive, wearable and implantable technologies. While novel technologies can affect our capabilities and our view of ourselves as individuals and as a species, I believe that some aspects of humanity will remain constant. Chief among them is the certainty that we will remain social beings, conscious of the possibility and reality of suffering, but also endowed with plans and hopes for a good life. We strive for happiness and seek to flourish in the knowledge that we will always be negotiating the question: how exactly can flourishing best be achieved? Technology can promote as well as reduce our flourishing. Our task is therefore to ask how novel technologies can affect flourishing and what we can do individually and collectively to steer such technologies in directions that support flourishing. I hope that this book will help us make positive use of AI and move towards a good, technology-enabled world.

Index

© The Author(s) 2021
B. C. Stahl, *Artificial Intelligence for a Better Future*,
SpringerBriefs in Research and Innovation Governance,
https://doi.org/10.1007/978-3-030-69978-9

Printed in the United States
by Baker & Taylor Publisher Services